Lecture Notes in Computer Science 1361

Edited by G. Goos, J. Hartmanis and J. van Leeuwen

Springer
Berlin
Heidelberg
New York
Barcelona
Budapest
Hong Kong
London
Milan
Paris
Santa Clara
Singapore
Tokyo

Bruce Christianson   Bruno Crispo
Mark Lomas   Michael Roe (Eds.)

Security Protocols

5th International Workshop
Paris, France, April 7-9, 1997
Proceedings

Springer

Volume Editors

Bruce Christianson
Computer Science Department, University of Hertfordshire
Hatfield, AL10 9AB, UK
E-mail: b.christianson@herts.ac.uk

Bruno Crispo
University of Cambridge, Computer Laboratory
New Museums Site, Pembroke Street, Cambridge, CB2 3QG, UK
and University of Turin, Department of Computer Science
Corso Svizzera 185, I-10149 Torino, Italy
E-mail: bruno.crispo@cl.cam.ac.uk

Mark Lomas
Information Security Department, Goldman Sachs International
Peterborough Court, 133 Fleet Street, London EC4A 2BB, UK
E-mail: mark.lomas@gs.com

Michael Roe
University of Cambridge, Computer Laboratory
New Museums Site, Pembroke Street, Cambridge, CB2 3QG, UK
E-mail: michael.roe@cl.cam.ac.uk

Cataloging-in-Publication data applied for

Die Deutsche Bibliothek - CIP-Einheitsaufnahme

Security protocols : 5th international workshop, Paris, France, April
7 - 9, 1997 ; proceedings / Mark Lomas et al. (ed.). - Berlin ;
Heidelberg ; New York ; Barcelona ; Budapest ; Hong Kong ;
London ; Milan ; Paris ; Santa Clara ; Singapore ; Tokyo : Springer,
1998
   (Lecture notes in computer science ; Vol. 1361)
   ISBN 3-540-64040-1

CR Subject Classification (1991): E.3, F.2.1-2, C.2, K.6.5, J.1

ISSN 0302-9743
ISBN 3-540-64040-1 Springer-Verlag Berlin Heidelberg New York

© Springer-Verlag Berlin Heidelberg 1998
Printed in Germany

Typesetting: Camera-ready by author
SPIN 10631798   06/3142 – 5 4 3 2 1 0   Printed on acid-free paper

Preface

Welcome to the proceedings of the fifth International Workshop on Security Protocols. These workshops grew from a series of informal meetings held at the University of Cambridge Computer Laboratory. Our aim has been to assemble researchers in an environment where they could discuss the limitations and omissions of current work in computer security, and the implications of these for future directions in security protocol research.

Since the publications in 1978 of the seminal paper on authentication by Roger Needham and Michael Schroeder, it has become abundantly clear that the properties which cryprographic protocols actually possess are extraordinarily fragile. One reason for this is the complex nature of the interactions between the algorithmic mechanisms used to realise the protocols on the one hand, and the high-level behaviour of the applications which the protocols are intended to support on the other. Experience also shows that it is difficult to abstract from these interactions successfully, and to describe them in a way which allows them to be reasoned about correctly.

Consequently, security failures often occur as a result of an unnoticed mismatch between the use an application makes of a security protocol and the properties which the realisation of the protocol provides.
The insights provided by these subtle constraints, and by breaking them, form the theme of this year's workshop. We hope these proceedings will enable you to share some of these insights.

We would like to thank Serge Vaudenay for the exemplary local arrengements at the Ecole Normale Superieure during the workshop.

October 1997   Mark Lomas
(Brumaire 206)   Bruce Christianson
                                     Bruno Crispo
                                    Michael Roe

Contents

Secure Books:
Protecting the Distribution of Knowledge

Ross J Anderson, Václav Matyáš Jr., Fabien A Petitcolas,
(Computer Laboratory, Cambridge CB2 3QG)
Iain E Buchan, Rudolf Hanka
(Medical Informatics Unit, Cambridge CB2 2SR)

University of Cambridge, UK
{rja14, vm206, fapp2, ieb21, rh10}@cam.ac.uk

Abstract. We undertook a project to secure the distribution of medical information using Wax. This is a proprietary hypertext-based system used for information such as treatment protocols, drug formularies, and teaching material. An initial attempt, using digital signatures (in line with a recent European standard) and certificates conforming to X.509 has thrown up a number of interesting problems with current approaches to public key infrastructures. While the X.509 philosophy may be suitable for many electronic commerce applications, signatures on which we may have to rely for many years — such as those on books and contracts — appear to require a different approach.

1   Introduction

Wax is a system for publishing electronic medical books containing information such as treatment protocols, drug formularies and government regulations to which healthcare professionals need frequent access in support of clinical decision-making. Its origins lie in an earlier project [5] and its primary goals were to provide a good enough clinician-computer interface to be used safely without much formal training, and to have the knowledge management structure needed to support local clinical practice.

Protection issues such as assuring the integrity of the information, the authenticity of its source and non-repudiation (in a broad sense) started to arise in late 1996. This resulted in a collaboration between two departments of Cambridge University: the Medical Informatics Unit (the developers of Wax) and the Computer Laboratory, which has a group interested in information security.

The protection priority is to ensure that Wax users can correctly identify the author and publisher of a book on which they rely for clinical decision making, both at the time and if need be in the event of a subsequent dispute. There is no requirement for strong secrecy properties; some books should be restricted to registered medical professionals only, but this is by virtue of drug adverts that may only be directed at this audience. So there is no more need to encrypt Wax books than to have the 'British Medical Journal' delivered by armed courier.

There is also no requirement to maintain an audit trail of which doctor or other healthcare professional read which chapters of which book. In fact, the maintenance of such a record would probably be considered an intrusion of professional privacy.

Thus Wax provides an interesting case study for the security professional. Unlike many other systems, whose protection is a mix of secrecy, authenticity, integrity and availability properties, Wax's needs are almost exclusively focussed on authenticity and integrity. It is also conceptually simple.

However, when we tried to implement protection mechanisms based on the obvious standards (X.509 [13] for certificates and the recently adopted European medical standard on RSA with exponent 3 for signatures) we ran into unexpected and interesting problems. These raise serious and important questions about the suitability of public key infrastructures currently under construction for certifying the integrity of long-lived objects such as books and contracts, and the authenticity of their signatories.

Section 2 overviews the Wax system. The threat model and security policy are discussed in section 3. The concept of trusted books and granularity of protection are discussed in section 4, while section 5 describes how the prototype system works and section 6 the trust structure. The problems encountered and lessons learned are described in section 7 and 8.

2   Wax

The original Wax project aimed to facilitate the sharing of knowledge between primary and secondary healthcare. The original system presented a user, such as a general practitioner, with a number of protocols for the care of specific diseases that were written and kept up to date by leading specialists in the field. Supporting information such as details of drugs were added, and in the next phase of the project it is planned to add administrative information such as directives from the Department of Health and local Health Authorities. The Wax mechanism also supports locally authored books, which might include policies and procedures developed by a general practice for its own use.

The educational experience of health care professionals is based around hierarchically classified paper systems such as libraries of books. Thus the source information for Wax is arranged into a familiar hierarchical structure: sections – books – shelves – library, and a special book-centred ASCII hypertext system was developed to support this.

The idea is that by providing a unified, hypertext-based library that enables clinicians to get at the information they need quickly and intuitively, and to update it when appropriate in a controlled manner. The system thus supports a clinical hypertext library appropriate to a healthcare provider such as a general practice or hospital. It also allows users to add their own notes to any topic of a Wax book (these notes are kept constant in a separate file to cope with book updates). A single book or the whole library can be searched for words or phrases, cross references between books can be made, and many books can be open at one time.

An example of the interface is shown in the following figure.

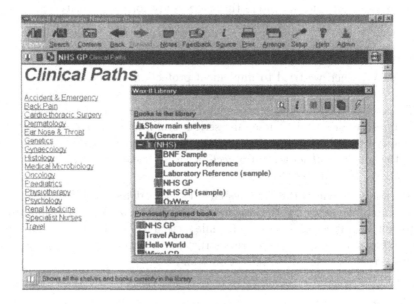

Figure 1. Wax Interface

The Wax software supports both browsing and authoring and has been optimised for clinical use on Microsoft Windows compatible platforms. Wax books can import and export HTML documents and are designed to be browsed mainly from local storage, as performance is critical to clinical usefulness. The added facilities which are not achievable using HTML and standard web browsers include feedback mechanisms, navigation logs and performance features. Wax software and relevant papers [5, 6] are at [http://www.medinfo.cam.ac.uk/wax/].

Wax book updates are done using diskettes or e-mail at present. However, from version 2 (due spring 1997) it is proposed to make updates available over the Internet by ftp or http. This could be particularly helpful in synchronising book updates between district level libraries and end-user. However, it naturally raised the question of what cryptographic security features might be advisable.

3 Threats and Countermeasures

The overall level of threat against Wax is low — certainly much lower than against systems involving identifiable personal health information. The following primary threats were identified:

1. a book's content could be altered, whether by accident or malice;

2. an incorrect book source (author or publisher) might be claimed;

3. Wax software could be maliciously altered, whether by a general virus or by a more targeted attack;

4. a party involved in a dispute might deny the content of a previously published book, or challenge the date on which the information was published.

The first three are familiar from the general computer security environment [2]; concern about the fourth arises from a case in which a supplier of surgical implants sought to defeat litigation by claiming that it had published warnings at an earlier date than it in fact had done. Thus in addition to the integrity and authenticity of books and software, we want a non-repudiation service that covers both content and publication dates.

Some of the attackers might be large companies involved in litigation and thus able and willing to use professional techniques. However, these would probably be targeting the non-repudiation mechanisms, and the great majority of attackers — and almost all of those in the first two threat categories above — are likely to be insiders; health-care professionals with limited computer experience. They might intend to change the information within a book, perhaps to cover up malpractice [1].

Although the third threat does not introduce a serious risk in our application, it is the main concern with distribution and installation. The issue facing us is where to draw 'the fine line between prudence and paranoia' when it comes to trusted distribution. The main control envisaged is that the second version of Wax will be distributed in CD form by post. The secondary control is that when Wax is installed and a user is registered, a hash of the installed software is printed on the registration form. The third level control is that we will make available several means of checking the Wax distribution's integrity, including a PGP signature and hashes published in the medical press.

Given these checks, a reasonable level of trust can be placed in the Wax software, and the master public key that it contains. This key can be manually verified at any time, and users are requested to check it against a published value on installation.

At installation, users are also requested to create a public-private keypair with which to their own communications may be signed and verified. Key generation is very similar to the procedure in PGP, except that a hash of the public key is printed on a registration form, together with the hash of the software mentioned above, and sent to the Wax-Centre by post. The public key is also sent to the Wax-Centre, preferably by email but if this is not possible by diskette or printed in hexadecimal on the registration form. After performing the appropriate due diligence checks (such as verifying medical registration), the Wax-Centre will send the user an identity certificate in the X.509 format.

Integrity checks are also performed whenever a book update is received, and whenever a book is opened (these will be described below, and are to a certain extent customisable by the user). The trust model is that the Wax-Root certifies the publisher, and the publisher certifies the book — taking responsibility for its content to the same extent as in the present world of paper (which is outside the scope of this paper).

The effect of the design is to reduce the problem of the trusted distribution of books to the trusted distribution of the Wax software and master public key. Under the circumstances, we consider that an appropriate level of effort has been expended on trusted distribution; any more effort than this would cross the line into paranoia.

4 Trusted and Untrusted Books

Each publisher certified by a Wax-Root certificate can publish books that the Wax system will consider to be trusted. At the present time, only Wax is a publisher, but other publishers can be added quickly, and in time most healthcare providers will act as publishers for their own local documentation. Users are permitted to have a small number of untrusted books open at any moment; these might be books that they are in the process of writing. However, once an author is finally satisfied with a book, a publisher can be asked to publish it, or the author can acquire the necessary key material for publishing. Once signed by a Wax-certified publisher, the book becomes trusted.

Each publisher is allocated a 'shelf' in the library, and users can clearly see the difference between trusted and untrusted books (different icons and colours). Users will also be warned if a book has been altered.

Users may determine how often books in their library are to be verified — ranging from when first downloaded to whenever opened. The reason for this is that delays must be minimised; the capability of user machines varies widely; and so does the threat environment in which the software is run, ranging from a single handed GP's notebook to a networked server in a large hospital with many temporary staff. So users need to select an appropriate frequency of checking.

Books can be verified in three different ways:

1. Each publisher maintains a catalogue that lists the currently available versions of all books, including not just their names but also their hash values. This catalogue is signed using the publisher's key, which in turn is certified using the Wax-Root key;
2. Each book is also signed by its publisher;
3. Version n of a book contains the hash of version $n-1$ (except for $n = 1$).

It is intended to provide a further level of non-repudiation — in view of possible attack by funded organisations involved in litigation — by depositing CDs containing the current Wax library at the UK's statutory copyright deposit libraries or a similar body. (The whole issue of whether electronic media should be subject to the same statutory copyright deposit rules is currently a matter of government consultation in the UK [7].)

A topic that we have still not fully resolved is the granularity of protection. A user with a slow computer (or following a slow hypertext link to an online book chapter) might not want to wait until the whole book has been verified. So there is a case for hashing each chapter of a book and then putting the top level protection on a hash of these hashes. Whether this brings more practical benefits than problems is to be explored empirically.

5   Book Updates

Whatever the internal granularity of protection, external protection (by the catalogue and signature mechanisms) is implemented at the book level. Thus each book has a hash value associated with it, that is protected both by publication in a publisher's catalogue and by being signed.

In order to issue an update of one or more books, a publisher must therefore sign them and also create either a complete new catalogue or a supplementary catalogue, which he also signs. This catalogue, plus the books it refers to, are made available for download.

The user first gets the catalogue and checks its signature all the way back through the certificate chain to the Wax-Root key. Books to be downloaded are then chosen, and Wax instructed to fetch them. Once the books have been downloaded and their integrity checked, an update is made to an index kept locally and it is signed with the user's signing key. A passphrase is solicited to confirm that the user is happy with the new configuration of the library.

6   Trust Structure

As noted above, the primary purpose of the Wax system's security features is to reduce the problem of verifying the authenticity and integrity of downloaded medical and other books to that of verifying the authenticity and integrity of the Wax software and its embedded root key. Only those books whose certificate chain can be followed successfully back to the Wax-Root are considered trusted.

The trust structure thus looks like figure 2 overleaf.

In the layer below the Wax-Root will be found the main publishers, such as the Wax-Centre (the current publisher); in time this may include both official publishers, such as the Department of Health, commercial publishers, and the publications of professional bodies such as the British Medical Association. As the number of publishers grows, it is envisaged that another one or two levels may be added. Thus, for example, the Department of Health may in time accept responsibility for certifying local health Authorities, who also publish; and the British Medical Association might certify individual general practitioners via the existing structure of Local Medical Committees. The future trust structure in medicine is a matter of current negotiation between interested parties, and

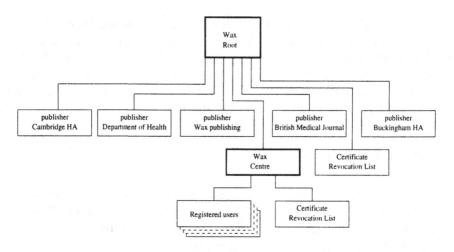

Figure 2. Foreseen Certification Structure

nothing we are doing in the Wax project is an attempt to pre-empt or second guess the outcome of these discussions.

But for the time being, to get the system underway, the Wax-Centre will act as the Certification Authority for individual users.

7 The Shortcomings of X.509 Certificates

In the initial design we produced for this system, we assumed that individual users and publishers would have X.509 certificates [13] as these are in some sense the 'standard' format (although there are competing architectures, such as Microsoft's SDSI [10]). The idea was that each user, and each publisher, would have a certificate, and that a Certificate Revocation list would be published daily as a special Wax book. However, once we got into the detailed design phase, it became increasingly apparent that the trust model implicit in X.509 is not particularly appropriate in electronic publishing.

This observation may be the principal value of this paper to the broader research community, and to people involved in building other applications, so it is worth going into in some detail.

X.509 certificates are in many ways similar to credit cards — certificates have an expiry date (like credit cards) and there is a certificate revocation list that performs the same function as the hot card list of a credit card company. Unsurprisingly, the support they offer for general purpose digital signatures is less than ideal.

Many signatures must persist for a long time, such as the signatures on Wax books (assuming for the time being that signatures continue to be used as a protection mechanism). What does it mean, for example, that an author's certificate expires after a set period of time? Does his book suddenly become untrusted? And what does revocation mean in the context of the signature on a book that has already been published?

'Planned obsolescence' may make sense in the world of software publishing, while in the world of electronic commerce, it is perfectly reasonable that the keys used to sign credit card transactions should be replaced every two years — just as physical credit cards are. This limits the period of time in which a live key remains open to compromise; it also limits the size of the certificate revocation list.

However, if a medical publisher goes to the trouble of having a textbook on a medical condition or procedure prepared by an eminent clinician, or where a drug company publishes the results of a clinical trial on a new drug, then we may wish that the integrity and usability of that information be protected for many decades. (The copyright in the work, for example, will typically persist until 70 years after the author's death.)

We are therefore faced with the problem of how we can assure the authenticity and integrity of Wax books for a long period of time, while at the same time not exposing signing keys to compromise by requiring that an individual maintain a single signing key, and migrate it from one system to another, for the whole of his life. (This is not just a medical publishing problem: other applications where durable and secure signatures are needed include not just archived medical records, but all kinds of legal contracts.)

We believe that the long term solution to the problem of long term trust will involve a thorough re-examination of many of the assumptions that have grown up around public key cryptography. We believe that catalogue mechanisms will provide a better way of protecting many long lived objects such as books and contracts; where an object needs to be authenticated and published more quickly than the catalogue update cycle will allow, and strong non-repudiation is a requirement, then a one-time signature may be ideal. After all, so long as a signing key still exists, so does the possibility that a compromise might occur and lead to problems with the validity of the signature.

So the typical two-to-three year lifetime of an X.509 certificate has the curious property that it is too short for long-lived trust, and too long for strong non-repudiation!

Other problems with X.509 in the medical publishing context include firstly that X.509 certificates vouch for identities rather than roles, while we need support for roles: notices signed by 'The Chief Medical Officer' should be bound with the office rather than with the individual who is presently its holder; and secondly that X.509 does not support dual control. There are of course workarounds: dual authorship can always be noted by having a book signed separately by its two authors, and a two-out-of-three arrangement for (say) the Wax-Root key can be achieved using threshold signatures. However, it would be much preferable to have explicit protocol level support for such features.

8 Catalogue Based Trust

The effect of these discoveries on the Wax project has been to persuade us to make the publisher's catalogue — rather than the signature of the publisher or the author — the primary check on integrity and authenticity as new books are downloaded; and the user's internal, signed, directory to be the main check as existing books are opened. The publishers' signatures on the books themselves are merely a secondary check.

So the primary function of the author's signature is now to vouch for the authenticity of his own list of installed books and thus ensure that important files have not been tampered with by other users. This is not entirely straightforward; another user could have replaced the list with one of his own, whether by accident or malice, and as this would be signed by the other user, it is not sufficient merely to check the certificate chain back to Wax-Root. This reduces essentially to the trusted path problem; we propose to display the certificate owner's name in the title of the list of books as well as in the title of every open book. Given that the Wax software is adequately trusted, this gives us an adequate level of confidence.

The secondary function of the user signing key is to validate books sent to a publisher. This is relatively unimportant (except with very prolific authors) as the author could always read a hash over the phone instead.

As an interim measure, we have decided to set a short lifetime (1 year) for the publishers' signing keys.

In the long term, we believe that trust in published matter will not rely on the certification of individual identity, but on cataloguing, notarisation and timestamping. In this model, each page may contain its own hash in a header, but will certainly contain the hashes of all trusted pages to which it points as an extension of the URL (these can be checked using a Java applet or other convenient mechanism). A publisher's root page will contain the top-level hash for his catalogue; the root page hash will be checkable by some other means (such as notarisation in a paper journal).

In this model, the main use of digital signature mechanisms may be to authenticate pages that are created on demand (such as current exchange rates), in which case mechanisms such as RSA may be used, and to provide flexible links to pages that are updated frequently (such as secondary care access information) or at short notice (such as an official warning about a drug side-effect). In the latter case, we suspect that on thorough analysis a prudent designer may well use one-time signatures, or at the very least conventional signature keys that are destroyed rapidly but whose corresponding verification keys remain valid in perpetuity.

This architecture is intrinsically congenial to web-based publishing and it may therefore succeed in the market place. In addition, it follows the Rossnagel principle that electronic trust structures should reflect those in existing practice [11] — a principle jointly agreed by the BMA and the Department of Health. If it does succeed — for whichever of these reasons — then the pragmatic approach being implemented in Wax will be as compatible with it as one can expect to be with standards that are not yet written. Our approach does however make such prudent use as can be made of the existing X.509 architecture that many governments and other organisations are struggling to put into the field; and if X.509 does come to underpin electronic publishing, then we are compatible with that too.

9 Conclusion

We have developed a mechanism to assure the authenticity and integrity of electronic books. Although our particular application was medical, many of the lessons learnt apply to publishing in general, to digital contracts, and indeed to any application in which we need to assure the trustworthiness of digital objects over long periods of time.

The main lesson learned was that the trust structure embodied in X.509 and related standards is not suited for such applications. Indeed, it may turn out that digital signatures are not the appropriate tool for the job, but rather secure cataloguing and notarisation services based on trees of hash values.

We managed to achieve a pragmatic compromise that enables us to move ahead with an initial solution that is broadly compatible with both the public key and cataloguing approaches. However, as time passes and more functionality is added, we expect that ultimately one or other of these design options will have to be closed off. We feel that for both market and other reasons, the protection of published matter will come to rely on mainly cataloguing, with digital signature techniques used principally to provide flexible links from a catalogue-based trust structure to items whose content varies more quickly than the catalogue update cycle.

A catalogue-based trust structure may assuage the fears of law enforcement agencies over crypto proliferation. In many applications (such as conventional book publishing) there is no need for secrets at all (except as part of local mechanisms for logon, trusted path and the like); even where digital signatures are used to link to urgent notices, there is every incentive to use one-time signatures, or at the very least signatures whose verification keys have very much longer lifetimes than their signing keys. In any case, the apparent requirement for a central identity authentication service evaporates.

In conclusion, the long term protection of the authenticity and integrity of digital objects is far from being an adequately solved problem.

Acknowledgements

Václav Matyáš Jr. would like to thank the Royal Society for supporting his research through a Postdoctoral Fellowship.

References

1. "Nurse sacked for altering records after baby's death", K Alderson, *The Times*, November 95, p 6
2. "Why Cryptosystems Fail", RJ Anderson, in *Communications of the ACM* vol 37 no 11 (November 1994) pp 32–40
3. "Security in Clinical Information Systems", RJ Anderson, published by the British Medical Association, January 1996
4. "The Eternity Service", RJ Anderson, *Pragocrypt 96*, proceedings published by CTU Publishing House, Prague, ISBN 80-01-01502-5, pp 242–252
5. "Decision Support for Primary Care Using the Path.Finder System", Iain Buchan, Heather Heathfield, Tom Kennedy, Peter Bundred, in *British Journal of Healthcare Computing v 13 n 6*, pp 20-22, July 1996
6. "Exchanging Clinical Knowledge via Internet", IE Buchan, R Hanka, in *MEDNET 96*, proceedings to be published as a CD-ROM
7. "Government plans to save e-media for sake of nation", *Computer Weekly*, February 20th 1997 p 16
8. Good Medical Practice, General Medical Council, UK
9. "GP Practice computer security survey", RA Pitchford, S Kay, *Journal of Informatics in Primary Care*, September 95, pp 6–12
10. "SDSI – A Simple Distributed Security Infrastructure", RL Rivest, B Lampson, at [http://theory.lcs.mit.edu/~rivest/publications.html], presented at USENIX 96 and CRYPTO 96, April 30, 1996
11. "Institutionell-organisatorische Gestaltung informationstechnischer Sicherungsinfrostrukturen", A Roßnagel, in *Datenschutz und Datensicherung (5/95) pp 259–269*
12. "Secure Hash Standard", National Institute of Standards and Technology, *NIST FIPS PUB 180*, U.S. Department of Commerce, May 1993
13. "Information technology – Open Systems Interconnection – The directory: Authentication framework", *ITU-T Recommendation X.509*, November 1993

Protocols Using Anonymous Connections: Mobile Applications

Michael G. Reed, Paul F. Syverson, and David M. Goldschlag
Naval Research Laboratory

Naval Research Laboratory, Center For High Assurance Computer Systems,
Washington, D.C. 20375-5337, USA, phone: +1 202.767.2389, fax: +1 202.404.7942,
e-mail: {*last name*}@itd.nrl.navy.mil.

Abstract. This paper describes security protocols that use anonymous channels as primitive, much in the way that key distribution protocols take encryption as primitive. This abstraction allows us to focus on high level anonymity goals of these protocols much as abstracting away from encryption clarifies and emphasizes high level security goals of key distribution protocols. The contributions of this paper are (1) a notation for describing such protocols, and (2) two protocols for location protected communication over a public infrastructure.

1 Introduction

As mobile devices for communication and computation gain more widespread acceptance, where a person is located when processing digital information or sending and receiving messages or phone calls is increasingly under individual control. Relatedly, individuals no longer tied to an office have enjoyed increasing privacy over their location information. If one can conduct business from anywhere, then one can be anywhere when conducting business. However, this is not an entirely accurate picture. For example, mobile phones may not reveal one's location to the party at the other end of the line as readily as stationary ones, but currently implemented technology still requires tracking of the mobile phone itself.

The primary purpose of phones is to allow individuals to communicate. Where anyone happens to be, and even who they are, is simply coincidental to that communication. Technology that more precisely reflects the functional needs of the intended application would therefore provide anonymous channels of communication. The communication over such a channel need not be anonymous; parties typically will identify themselves over the channel, but the channel itself should not reveal their locations or identities to the network or observers of the network. This paper describes protocols that use anonymous channels as primitives. After sketching the requirements for a channel to be anonymous we use such channels to construct protocols for location protected mobile applications. One such application we have already mentioned. Specifically, our protocol allows a mobile phone to send and receive calls without revealing its location to anyone, including the communications infrastructure on which it relies. A side benefit of our protocol is that its implementation would potentially extend the useful battery life of a phone in standby (listening) mode by orders of magnitude.

Another, less well known but increasingly important, application of mobile communication is in location tracking. This has already been implemented in the Lo-Jack system that allows police to track a car that has been reported stolen and in an active badge system implemented by Olivetti at the Computing Laboratory at Cambridge University. It is also an important component of ITS (intelligent transport systems, cf., below). As an example we describe active badge systems. Like current mobile phones, these do not fully protect location information. In fact quite the opposite. The purpose of such a system is to track the location of those wearing the badges. This can be useful, e.g., in an environment where individuals are not always in their offices but it is important to be able to find them when needed. While useful, active badges can have overtones of big brother, as they allow a company to track things such as how long an individual is in the cafeteria. One way to reduce this threat is to give individuals control over their location information. Such concerns have led to the proposal of protocols for doing just that [7]. These protocols allow individuals to keep their location information in a designated repository over which they have access control. Not even the tracking system is able to determine where an individual is without consulting the designated repository. This paper also presents protocols for individual control of location information in a tracking system. Our approach is related to that of Jackson in [7] but has important differences. One difference is that all versions proposed in [7] require the badge to produce or carry route information for each message it sends. Our approach requires only the production of an adequately random string of one-time identifiers. This means that the Jackson approach decentralizes the control of location information, which appears to be good, and makes the badge operation more complicated, which appears to be bad. Our approach allows for simple badges but requires a centralized database. This centralization might appear to be a vulnerability, but we shall see that it is not.

Intelligent transport systems are designed to track the movement of vehicles on appropriately structured public highways. Some of the advantages of such a system include route optimization for individual vehicles, traffic control for all vehicles so enabled, and traffic signal control for emergency vehicles. However, there is great potential for abuse in such a system [8]. In addition to civil rights and privacy abuses similar to the problems described above for badging systems, the potential exists for other, perhaps more serious threats. For example, truck hijackers could make use of a system that tracks the movement of a fleet of trucks to optimize their chances for a successful and lucrative hijacking. Kidnappers, terrorists, murderers, etc. could trace someone, even if, e.g., she intentionally took varying routes to work each day.

While the private location tracking protocol described in this paper is explained in terms of the active badging example, it is a general protocol for private location tracking using a public infrastructure. Thus, it applies equally well to vehicle tracking in ITS or an enhanced Lo-Jack system. In fact, this protocol is a special case of location protected communication using a public infrastructure, where what is communicated is itself location information.

The remainder of the paper is structured as follows: In section 2 we give an overview of anonymous channels and present our notation for describing protocols that make use of them. In section 3 we set out two protocols, a protocol for location protected communication over cellular phones in section 3.1 and a protocol for private location tracking in section 3.2. In section 4 we present background information. In particular, we briefly describe onion routing, a system we have implemented for anonymous communication over the Internet. In section 5 we present our conclusions.

2 Anonymous Channels

For us, an *anonymous channel* is a communication channel for which it is infeasible to determine both endpoints. The principal initiating the connection is the *initiator*, and the principal to whom he connects is the *responder*. These are not merely theoretical constructs; we have implemented a mechanism for anonymous channels (in fact near real-time anonymous connections) that operates below the application layer and supports a variety of Internet applications [10, 11]. The design of our mechanism can be applied to non-Internet applications such as are described in this paper. We will give more background on our design in section 4.1. Just as the strength of an encryption algorithm is relative to assumptions about everything from special restrictions on the key space and on other inputs to the capabilities and collateral information of a potential attacker, so too what we mean by "infeasible" will have many caveats and limitations for any given mechanism to implement an anonymous channel. Fortunately, for protocol purposes we need not specify these any more than we need to specify properties of cryptographic algorithms and their implementations when we describe, for example, a protocol for authenticated key distribution.

In practice, we do assume that the initiator knows the responder: since he initiates the connection, he presumably knows to whom he intends to connect. Note that this assumes either a means to authenticate the responder or, in the case of one way communication, a means to guarantee that only the responder can receive the message, e.g., public key encryption. In theory, the initiator may be sending out a proverbial note in a bottle, destined for where he does not know. However, unlike a shipwrecked sailor on a desert island, our initiator will have reason to believe that his messages can ordinarily be tracked back to him the moment he releases them. Therefore, he will need to assure himself that his messages have drifted far enough away from him before anyone can begin to track them. Thus, even if he wishes to establish an anonymous channel with whomever will respond, he needs to determine a point away from himself before which the channel will not emerge.

The initiator may build an anonymous connection all the way to the responder. This would protect both of them from association with the channel by all but the initiator. However, since the initiator often needs only to hide that communication is coming from or going to him, in practice we may only have half-anonymous channels. In other words, the initiator produces a channel which cannot be traced to him and uses this to contact the responder. From the end of the anonymous part of the channel to the responder anyone can see what the responder is sending and receiving. If end-to-end encryption is piped through this half-anonymous channel, it can effectively be made fully anonymous. Nobody can tell what the responder is sending or receiving or to whom the responder is connected. The only thing that can be observed from the outside is that the responder is talking to someone. (This too can be hidden if the channel is maintained even when not in use and dummy traffic sent over it. However, such countermeasures are quite expensive.)

Note that anonymous channels are not explicitly required to be confidential channels. However, cleartext is obviously trackable. And, even ciphertext that appears the same everywhere is trackable. Thus, for the reasons we have been describing, an anonymous channel must be encrypted in a changing manner at least to a point where it is acceptable that the communication be tracked.

Before going further, we contrast anonymous channels with a related but distinct form of channels, specifically subliminal or covert channels. In theory, all of these are channels for which it is infeasible to detect the existence of the channel. Thus, our distinction deals more with the expected environment for a channel rather than the channel's undetectability in that environment. In practice, channels called 'subliminal' typically piggyback on legitimate channels between the principals. Covert channels either piggyback similarly or exist in a medium that is not explicitly a communications medium at all. So, covert and subliminal channels are channels that rely on some other type of channel or computation to hide them. (In most applications the covert or subliminal channel will be somehow illegitimate and the cover channel or computation legitimate.) By contrast, anonymous channels rely on other anonymous channels to hide them. The cover for these channels are other channels like them, and the hiding comes from the indistinguishability of these channels from each other. Another important contrast between these types of channels is their relative efficiency. Though not always the case, covert and subliminal channels are typically inefficient as compared to the legitimate communication they parallel. Anonymous channels are expected to be roughly comparable in efficiency to their non-anonymous counterparts in the same medium. We have found this to be the case in our implementation.

2.1 Anonymous Connections

The channels we consider for these applications are connection based. Thus, we will typically speak of anonymous connections rather than anonymous channels in general. We denote the sending of message M along an anonymous connection from X to Y by '$X \Rightarrow_X Y : M$'. It may be important to know if a message is being sent over an anonymous connection from initiator to responder or vice versa. Specifically, it may be important to know whose identity is protected from association with the channel. This is the purpose of the subscript in the just introduced notation. If Y sends X a message M' on an anonymous channel that X initiated, this is denoted '$Y \Rightarrow_X X : M'$'. Sending M on an ordinary (non-anonymous) connection from X to Y is denoted '$X \rightarrow Y : M$'.

Observation: $X \Rightarrow_X Y \rightarrow Z : M$ implies $X \Rightarrow_X Z : M$.

2.2 Replies to Anonymous Connections

It is also possible for the initiator to make available information that allows a specific or arbitrary responder to establish a connection back to the initiator. This connection will be anonymous just as if it were a connection established from the initiator. We call such a connection a reply-to-anonymous (RTA) connection. The data structure that allows a principal to make an RTA connection to X is denoted '$\langle \Rightarrow_X X \rangle$'. Note that if the responder builds the RTA connection on the end of an anonymous connection in which he is the initiator, the result is a connection in which neither principal can be identified (unless he sends identifying information through the connection).

3 Mobile Applications

We now consider some applications of anonymous channels. Specifically, we describe how anonymous connections may be used to hide location information in cellular phone and location tracking systems.

3.1 Location Hiding for Cellular Phones

First we will describe how to make calls to a cellular phone without requiring the phone to reveal its location. Then we will describe how to hide the location of a caller's cellular phone from the both the network and external eavesdroppers. There are other solutions with similar anonymity goals that contain many of the elements in our protocol [2, 3]. Our protocol focuses on simple yet anonymous communication on top of an energy efficient call-back architecture.

In current cellular phone systems, the location of a phone is tracked, so calls to that phone can be routed through the base station controlling the phone's current cell. This tracking has two disadvantages: One is that the system knows where phones are. The other is that phones must transmit frequently to update their locations. This drains the phone's battery quickly.

In our proposal, instead of tracking a phone's location, phones will be paged. When such a phone is called, the network invents a temporary number and pages the phone. The phone's response to the page will be to make a call back to the temporary number in the page. The phone network will then mate the two connections. In addition to overcoming the disadvantages just mentioned, this simplifies our protocol because we effectively need to describe only how to initiate a call from a cell phone; the phone never receives a call in the ordinary sense of 'receive'. We will return to discuss paging briefly below.

The principals specified in our protocol are the caller's cell phone P, the central switch S, and the callee intended to receive the call R. We now present our protocol for initiating a call from a cell phone.

1. $P \Rightarrow_P S$: *Payment info.*, N
2. $S \Rightarrow_P P$: *Ack* or *Nack*
3. $P \Leftrightarrow_P S \leftrightarrow R$: *Conversation*

To make a call from a cellular phone without revealing location, the phone makes an anonymous connection to a central switch. It then sends to the switch the number it is trying to reach, together with some payment information to cover billing. The payment information may be the phone's subscriber ID or a credit card number or even anonymous e-cash of some sort. N is either R's phone number in the outgoing case or the temporary number from the page in the incoming case. Assuming that the payment information is acceptable the switch allows the call to be completed. In the outgoing case, the switch completes the call to R and patches this to the anonymous connection from P. In the incoming case, the switch allows the the connection from R to be patched to the anonymous connection from P.

Since we are only trying to hide location, the anonymous connection need not be made all the way from the caller to the callee. Rather, the anonymous connection is made to some central switch in the network, from which it can be passed along in the clear. This switch will not know from where the call is coming; however, it will not complete the call unless the phone sends identifying information or some guarantee of payment. (We do not here discuss how identification is authenticated.)

There is nothing in the protocol description to indicate that we are dealing with mobile phones. Of course with stationary phones, location protection of a given phone is moot. But, the protocol still protects the location origin of a call made from that phone. In fact, this sort of protection for stationary phones is discussed in [9]. It is helpful to have a notation abstract enough to cover anonymity in both these stationary phone connections and the mobile connections of [2, 3]. This is as it should be because the means to establish anonymous connections is separate from the basic communication medium that underlies it. For example, in our Internet implementation (cf. section 4.1) the underlying network is free to make whatever dynamic routing choices between points that it ordinarily does, provided that it connects to the points we do specify. Thus, the usual mobile phone procedures for connection to local base station and hand-offs between base stations as a phone changes cells are unaffected by our anonymous connections. The fact that there is movement in the cellular phone network is not hidden from the network; however, who is talking and where they are is hidden. So, the network is untrusted in this sense.

Our combination of anonymous connections and paging has two advantages. The locations of inactive phones do not need to be tracked within a paging region. Also phones never need to transmit except when they are involved in a call. This greatly reduces battery drain. For example, pagers last for a few months on a single battery, while cell phones last about a day in standby mode.

Many people would like to carry a cell phone to call or be called only in emergencies. Right now this is only convenient for outgoing calls. For incoming calls this is tedious at best since it still requires virtually daily charges of the battery. Our combination would make this more feasible since the phone could be carried for a month or more without recharging. It also has advantages over carrying a pager and a switched off cell phone. Aside from the advantage of needing to carry only one small device, calls from stations that do not allow incoming calls, e.g., payphones, can be taken.

One could imagine a variety of subscription prices for incoming calls based on the type of paging that is made available. Basic pagers typically operate in a large region (relative to cell phone cells), but basic service will not cover a large country like the US. Someone who regularly travels nationally might opt for a more expensive national (or international) paging service. In between these two extremes, one could have a phone-pager that operates regionally but updates the paging region it is in the first time it is turned on in that region. (Once it changes regions, it cannot receive incoming calls until it updates the region.)

3.2   Private Location Tracking

The next application we discuss is a location tracking service for which the user can control access to his location information. An active badging system can provide location information for individuals by sending badge identifiers to room sensors. Such a system, for example, has been implemented by Olivetti at the Cambridge University Computing Laboratory. While this information is useful for tracking people down, it may be a little too useful, making people hesitant to use it willingly. If control over access to an individual's location information can be placed in the hands of that individual, the system becomes much less threatening. The goal is then to provide a trusted home machine that can track the location of its user without centralized tracking information arising anywhere else and with limited computation power necessary for the wearable tracking device.

Here is a basic description of such a system. In terms of computing power of the wearable device (badge), it requires only that the user's wearable device share a pseudo-random number generator (PRNG) with the user's home machine. The PRNG is used to produce a sequence of tags that will serve roughly as one-time passwords. The badge sends a new tag every time it detects a different sensor. Once it receives an acknowledgement from the sensor it advances to the next tag. The badge must assume that the tag will be sent to the home machine at that point. The sensor opens an anonymous connection to the database and sends the current tag encrypted for the database and sends a (symmetric) key for encrypting the reply. The database looks up the RTA data structure associated with that tag (which the home machine has deposited there) and sends it back to the sensor. The sensor then uses this to construct an RTA connection to the home machine and sends the home machine his name (i.e., location information) and the tag. The principals specified in our protocol are: the wearable badge B, the room sensor S, the central database D, and the user's home machine H.

1. B detects new S
2. $B \rightarrow S : Tag$
3. $S \rightarrow B : Ack$
4. $S \Rightarrow_S D : \{Tag, K_{ds}\}_{K_d}$
5. $D \Rightarrow_S S : \{\langle \Rightarrow_H H \rangle\}_{K_{ds}}$ $(\langle \Rightarrow_H H \rangle$ stored at D under Tag)
6. $S \Rightarrow_H H : S, Tag$

A few assumptions are necessary. We assume that the database is query only. This prevents an attacker from reading random RTA data structures from the database and confusing the home machine about the whereabouts of the badge. Even if attackers could read random RTAs, the attack would not reveal any information (but might confuse the home machine with bogus tags). Specifically, this attack would not allow an attacker to frame a badge wearer by sending a sequence of bogus locations to the home machine since there is no way to identify successive tags or matching RTA structures. Despite this, requiring that the database be query only will mean that the only way to mount such an attack would be to guess tags or to grab them from a badge as it sends them out, presumably requiring hardware and a more concerted attack. The home machine is assumed to have deposited (over an anonymous connection) an RTA structure for each tag. This should not be done in a batch unless the user wants to trust the database to know that all of the deposited tags and RTA structures are from the same home machine. However, many RTA structures can be deposited in advance. In fact, depositing several RTAs in advance makes this protocol resistant to attack via spurious Acks. Since each Ack makes the badge move to the next tag, spurious Acks might cause the badge and home station to drift out of synch. More specifically a query to the central database may not find a matching RTA. However, submitting enough RTAs in advance solves the spurious Ack problem, provided that the home machine checks ahead whenever it receives a tag that does not match the next one expected. The protocol is also resilient to lost Acks, even without depositing more than one RTA in advance. If the badge does not receive a sent Ack but the matching RTA is used, then only the next location update is lost.

There are some vulnerabilities associated with this protocol. Corrupt sensors could fail to acknowledge receipt of the tag. They could then cooperate to track the user who would be sending the same tags as it moved about. This attack is limited in that the tags will change every time the device encounters a properly functioning and uncompromised sensor. The device could also keep track of the number of times (or number of successive times) it fails to receive an Ack. If a threshold number is exceeded it could either cease to operate or flash or beep to indicate an error. This is useful for more than prevention of attacks since failure to properly receive $Acks$ could indicate a malfunction in the wearable device or in the locating system. Note that corrupt sensors cannot otherwise cooperate with each other or with a corrupt centralized database since there is nothing to correlate successive connections to the database. (If a badge moves from the range of one corrupt sensor to another, they could of course observe the successive connections made. However, even then it might take a good deal of analysis to determine which badge is likely to have followed which path, especially if a user passes through rooms with several other individuals.)

An alternative that would prevent the above attack is for a badge to send out a new tag at regular intervals, whether it encounters a new sensor or not. This probably entails more overhead since individuals are likely to spend extended intervals at given locations, e.g., in their offices. Notice that the database cannot infer that someone is sedentary (much less who) because there is no way to link successive queries as coming from or not coming from the same badge. And, the connection to it from the sensors is always anonymous, so it cannot tell whence queries come.

Another alternative protocol avoids the use of anonymous connections entirely. Here there is no centralized database. Instead sensors simply broadcast tags (and sensor IDs) to all home machines. Home machines then pick up those tags that are meant for them to track their users. There are a variety of tradeoffs between this protocol and those that make use of anonymous connections, e.g., cost of anonymous connection set-up vs. cost of broadcast. Which is the better approach is likely to be highly contextual.

4 Background

Chaum [1] defines a layered object that routes data through intermediate nodes, called *mixes*. These intermediate nodes may reorder, delay, and pad traffic to complicate traffic analysis. Chaum's mixes and related work are the basis for almost all subsequent work on anonymous communication. Other approaches to anonymity in mobile phone systems occur in [2] and [3]. Another approach to private location tracking occurs in [7]. The approach to anonymous connections that we have implemented is called *onion routing*. Onion routing shares many anonymity mechanisms with Babel [6] but Babel uses them specifically for e-mail, while onion routing uses them to build (possibly long lived) application independent connections. We now give a basic description of onion routing; more details can be found in [4, 10, 11, 5].

4.1 Onion Routing

Traffic analysis can be used to infer who is talking to whom over a public network. For example, in a packet switched network like the Internet, packets have a header used for routing, and a payload that carries the data. The header, which must be visible to the network (and to observers of the network), reveals the source and destination of the packet. Even if the header were obscured in some way, the packet could still be tracked as it moves through the network. Encrypting the payload is similarly ineffective, because the goal of traffic analysis is to identify who is talking to whom and not (to identify directly) the content of that conversation.

Onion routing protects against traffic analysis attacks from both the network and observers. Onion routing works in the following way: The initiating application, instead of making a connection directly to a responding server, makes a connection to an application specific *onion routing proxy*. That onion routing proxy builds an anonymous connection through several other *onion routers* to the destination. Each onion router can only identify adjacent onion routers along the route. When the connection is broken, even this limited information about the connection is cleared at each onion router. Data passed along the anonymous connection appears different at and to each onion router, so data cannot be tracked en route and compromised onion routers cannot cooperate. An onion routing network can exist in several configurations that permit efficient use by both large institutions and individuals. The onion routing proxy defines a route through the onion routing network by constructing a layered data structure called an *onion* and sending that onion through the onion routing network. Each layer of the onion is public key encrypted for the intended onion router and defines the next hop in a route. An onion router that receives an onion peels off its layer, reads from that layer the name of the next hop and the cryptographic information associated with its hop in the anonymous connection, pads the embedded onion to some constant size, and sends the padded onion to the next onion router.

Before sending data over an anonymous connection, the initiator's onion routing proxy adds a layer of encryption for each onion router in the route. As data moves through the anonymous connection, each onion router removes one layer of encryption, so it finally arrives as plaintext. The last onion router forwards data to another type of proxy, called the *responder's proxy*, whose job is to pass data between the onion network and the responding server. This layering occurs in the reverse order for data moving back to the initiator. So data that has passed backward through the anonymous connection must be repeatedly decrypted to obtain the plaintext.

5 Conclusion

This paper presents a means for talking about anonymous connections and protocols that make use of them. We demonstrate the usefulness of describing anonymous connections at this level of abstraction by using our notation to describe two protocols for location protected communication over a public infrastructure.

References

1. D. Chaum. Untraceable Electronic Mail, Return Addresses, and Digital Pseudonyms, *Communications of the ACM*, v. 24, n. 2, Feb. 1981, pages 84-88.

2. H. Federrath, A. Jerichow, D. Kesdogan,and A. Pfitzmann. Security in Public Mobile Communication Networks, *Proceedings of the IFIP TC 6 International Workshop on Personal Wireless Communications*, Verlag der Augustinus Buchhandlung Aachen, 1995, pages 105–116.

3. H. Federrath, A. Jerichow, and A. Pfitzmann. MIXes in Mobile Communication Systems: Location Management with Privacy, in *Information Hiding*, Ross Anderson ed., Springer-Verlag, LNCS vol. 1174, June 1996, pages 121–135.

4. D. Goldschlag, M. Reed, and P. Syverson. Privacy on the Internet, *INET '97*, Kuala Lumpur, Malaysia, June, 1997.

5. D. Goldschlag, M. Reed, and P. Syverson. Hiding Routing Information, in *Information Hiding*, Ross Anderson ed., Springer-Verlag, LNCS vol. 1174, June 1996, pages 137–150.

6. C. Gülcü and G. Tsudik. Mixing Email with *Babel, 1996 Symposium on Network and Distributed System Security*, San Diego, February 1996.

7. I. Jackson. Anonymous Addresses and Confidentiality of Location, in *Information Hiding*, Ross Anderson ed., Springer-Verlag, LNCS vol. 1174, June 1996, pages 115–120.

8. P. Karger and Y. Frankel. Security and Privacy Threats to ITS, *The Second World Congress on Intelligent Transportation Systems*, Yokohama, Japan, November 1995, pages 2452–2458.

9. A. Pfitzmann, B. Pfitzmann, and M. Waidner. ISDN-Mixes: Untraceable Communication with Very Small Bandwidth Overhead, *GI/ITG Conference: Communication in Distributed Systems*, Mannheim Feb, 1991, Informatik-Fachberichte 267, Springer-Verlag, Heildelberg 1991, pages 451-463.

10. M. Reed, P. Syverson, D. Goldschlag. Proxies for Anonymous Routing, *Proceedings of the 12th Annual Computer Security Applications Conference*, IEEE CS Press, December, 1996, pages 95–104.

11. P. Syverson, D. Goldschlag, and M. Reed. Anonymous Connections and Onion Routing, *Proceedings of the Symposium on Security and Privacy*, Oakland, CA, May 1997, pages 44–54.

Receipt-Free Electronic Voting Schemes for Large Scale Elections

Tatsuaki Okamoto

NTT Laboratories
Nippon Telegraph and Telephone Corporation
1-1 Hikarinooka, Yokosuka-shi, Kanagawa-ken, 239 Japan
Email: okamoto@sucaba.isl.ntt.co.jp
Tel: +81-468-59-2511
Fax: +81-468-59-3858

Abstract

This paper proposes practical receipt-free voting schemes which are suitable for (nation wide) large scale elections. One of the proposed scheme requires the help of the voting commission, and needs a physical assumption, the existence of an untappable channel. The other scheme does not require the help of the commission, but needs a stronger physical assumption, the existence of a voting booth. We define receipt-freeness, and prove that the proposed schemes satisfy receipt-freeness under such physical assumptions.

1 Introduction

Various types of electronic secret voting schemes have been proposed in the last ten years [BGW88, BT94, CCD88, CFSY96, Cha88, FOO92, GMW87, Ive92, JSI96, Oka96, SK94, SK95], and recently *receipt-free* voting schemes are attracting many researchers [BT94, JSI96, Oka96, SK95]. The receipt-free property means that voting system generates no receipt (evidence) of whom a voter voted for, where the receipt of a vote, which proves that a voter has voted for a candidate, could be used by another party to coerce the voter.

Benaloh and Tuinsra [BT94] introduced the concept of the receipt-free voting based on the framework of the voting scheme using higher degree residue encryption [BY86, CF85]. They used a physical assumption, the existence of a *voting booth*. Their scheme allows voters only yes/no voting and is very impractical for large scale elections, since a lot of communication and computation overhead is needed to prevent the dishonesty of voters by using zero-knowledge (like) protocols.

Sako and Kilian [SK94] and Cramer, Franklin, Schoenmaker and Yung [CFSY96] improved the efficiency of the underlying zero-knowledge protocols by using discrete logarithm encryption in place of the higher degree residue encryption used in [BY86, CF85, BT94]. However, their schemes do not satisfy receipt-freeness. Moreover, their scheme allows voters only yes/no voting, and if it is extended to multiple bit voting, their schemes are still inefficient in practice.

Sako and Kilian [SK95] proposed a receipt-free voting scheme based on the Mixnet framework [Cha81]. Their scheme uses a weaker physical assumption, the existence of an *untappable channel*, than the physical assumption, a voting booth, of [BT94]. Their solution also satisfies universal verifiability. However, their scheme allows voters only yes/no voting, and if it is extended to multiple bit voting, their scheme is very inefficient in practice, especially when it is used for a large scale voting system.

Here, an *untappable channel* for V is a physical apparatus by which only voter V can send a message to a party, and the message is perfectly secret to all other parties. A *voting booth* is a physical apparatus for V in which only voter V can interactively communicate with a party, and the communication is perfectly secret to all other parties.

Another practical approach for realizing electronic voting involves the schemes using blind signatures and anonymous channels [Cha88, FOO92, Oka96]. This approach is considered to be the most suitable and promising for large scale elections, since the communication and computation overhead is fairly small even if the number of voters is large. Moreover, this type of scheme naturally realizes multiple value voting, and is also very compatible with the framework of existing physical voting systems.

In addition, this type of scheme is universally acceptable, and this is the most important property in election systems, since otherwise many people should be suspicious about the voting result. We now explain the reason why this framework is universally acceptable. The procedures consist of four stages; the authorizing stage, voting stage, claiming stage, and counting stage. In the authorizing stage, the administrators issue blind signatures. In the voting stage, the voters send their votes with the administrator's signatures to the bulletin board (or counter) through anonymous channels. In the claiming stage, each voter can publicly claim if his/her vote is not found in the board, and in the counting stage, the votes on the board are verified and counted. Here, in the claiming stage, everyone has the chance to raise a claim if he/she is suspicious about the contents of the board, and anyone (e.g., judge) can clearly determine whether the claim is valid or not, by checking the validity of the administrator's signature included in the claim. Thus, at the end of the claiming stage, everyone should be satisfied with the contents of the board (otherwise he/she should have raised a claim and had it resolved), and should be satisfied with the voting result, since all can count the voting result from the contents of the board.

[Oka96] proposed a *receipt-free* voting scheme based on this framework. To our best knowledge, this scheme is the only receipt-free voting scheme that is based on this framework and is considered to be practical for large scale elections.

However, in this paper, we show a security flaw in the receipt-free property of this scheme, and propose some new voting schemes to overcome this security flaw. One scheme requires the help of a group of the voting commission, called the "parameter registration committee" (PRC), and needs the physical assumption of an *untappable channel*. Another scheme does not require the help of such a committee, but needs the stronger physical assumption of a *voting booth*. Since both solutions are still practical, the proposed receipt-free voting schemes are suitable for practical (nation wide) large scale elections.

One of the reasons why [Oka96] had such a flaw in receipt-freeness is that no formal definition and proof of receipt-freeness have been given in [Oka96]. Although Benaloh and Tuinstra [BT94] have defined receipt-freeness, their definition is specific to their framework, and cannot be used in our framework. Therefore, it is very important to define receipt-freeness based on our framework, and to prove that a voting scheme satisfies this definition.

This paper defines receipt-freeness based on our framework, and proves that our modified schemes satisfy receipt-freeness under physical assumptions (i.e., an untappable channel or a voting booth).

This paper is organized as follows: Section 2 introduces the previous voting scheme [Oka96], Section 3 shows a security flaw in [Oka96], and Section 4 gives the definition of receipt-freeness. In sections 5, 6 and 7, our voting schemes are presented and are proven to be receipt-free under physical assumptions.

2 Brief description of the previous scheme

This section briefly introduces [Oka96].

2.1 Participants of the proposed scheme

The participants of this scheme are voters, V_i ($i = 1, 2, \ldots, I$), and voting commission, which consists of multiple administrators, multiple privacy commission members, and multiple timeliness commission members. Note that this scheme assumes no anonymous channel through the use of the Mixnet method [Cha81] with the multiple privacy commission members.

However, to simplify the explanation of this scheme, hereafter we assume that the voting commission consists of a single administrator, A, and a single timeliness commission member, T. In addition, we assume an anonymous channel but no privacy commission members.

2.2 Procedures

[Authorizing stage]

Several parameters, p, q, g, h, are generated and published by the system, where p and q are prime, $q|p-1$, g and h are in Z_p^*, and $q = \text{order}(g) = \text{order}(h)$. Here, α such that $h = g^\alpha \mod p$ is not known to any party.

1. V_i randomly generates $\alpha_i \in Z_q$, and calculates $G_i = g^{\alpha_i} \mod p$. We then define $BC(v_i, r_i) = g^{v_i} G_i^{r_i} \mod p$. Here, $BC(v_i, r_i)$ is a *trap-door bit-commitment*, since V_i can open this bit-commitment in many ways, (v_i, r_i), (v_i', r_i'), etc., using α_i such that $v_i + \alpha_i r_i \equiv v_i' + \alpha_i r_i' \pmod{q}$.
   V_i makes his/her vote v_i and computes

   $$m_i = BC'(v_i, r_i) = g^{v_i} G_i^{r_i} \mod p,$$

   using random number r_i. V_i computes

   $$x_i = H(m_i \| G_i) t_i^e \mod n,$$

   where t_i is a random number in Z_n, and (e, n) is the RSA public key of A for signatures, and H is a hash function. (x_i is a blind message for the RSA blind signature.) V_i generates his/her signature $z_i = S_{V_i}(x_i)$ for x_i. V_i also computes

   $$E_A(x_i \| z_i \| ID_{V_i}),$$

   where E_A is public-key encryption using A's public-key, and $\|$ denotes concatenation.
2. V_i sends $E_A(x_i \| z_i \| ID_{V_i})$ to A.
3. A decrypts the message, and checks that voter V_i has the right to vote, by using the voters' list. A also checks whether or not V_i has already applied. If V_i doesn't have the right or V_i has already applied, A rejects. If V_i is accepted, A checks the signature z_i of message x_i. If they are valid, then A generates signature

   $$y_i = x_i^{1/e} \mod n.$$

   A sends y_i to V_i.
4. V_i gets A's signature $s_i = H(m_i \| G_i)^{1/e} \mod n$ of message m_i by $s_i = y_i / t_i \mod n$ (i.e., unblinding procedure).

[Voting stage]

V_i sends $(m_i \| G_i, s_i)$ to the bulletin board through an anonymous channel. V_i also sends (v_i, r_i, m_i) to timeliness commission member T through an untappable anonymous channel.

[Claiming stage]

V_i checks that his/her ballot is listed on the bulletin board (ballot list). If his/her vote is not listed, then V_i claims this by showing $(m_i \| G_i, s_i)$.

[Counting stage]

In this stage, T publishes the list of votes, v_i, in random order on the board, and also shows a non-interactive modification of zero-knowledge proof, σ, to prove that the list of v_i contains only correct open values of the list of m_i without revealing the linkage between m_i and v_i. In other words, T publishes $(v'_1, \ldots, v'_I)$, which is a random order list of v_i. That is, $v'_i = v_{\pi(i)}$ $(i = 1, \ldots, I)$, where π is a random permutation of I elements. Given $(m_1, \ldots, m_I)$ and $(v'_1, \ldots, v'_I)$, T proves that T knows (π, r_i) such that

$$m_i = BC(v_i, r_i), \quad v'_i = v_{\pi(i)},$$

without revealing (π, r_i).

Here, we omit the description of how to calculate σ.

3 A security flaw in the receipt-freeness of the scheme

In [Oka96], the trapdoor bit-commitment is essential for satisfying receipt-freeness. If the value of α_i is generated by voter V_i as specified, then the scheme satisfies the receipt-freeness.

However, if α_i is generated by a coercer C, and C forces V_i to use $G_i = g^{\alpha_i} \bmod p$ for V_i's bit-commitment, then V_i cannot open $m_i = BC(v_i, r_i)$ in more than one way, since V_i does not know α_i. Hence, the voting scheme is not receipt-free and C can coerce V_i. (Here, we assume that C will pay V_i money or release a hostage, if C gets the receipt indicating that V_i voted in C's favor.)

4 Definition of receipt-freeness

This section defines the receipt-freeness based on the above-mentioned framework of voting schemes.

Definition 4.1 *Given published information, X, (public parameters and information on the bulletin board), adversary (coercer) C interactively communicates with voter V_i in order to force V_i to cast C's favorite vote v_i^* to T, and finally C decides whether to accept $View_C(X : V_i)$ or not, and T decides whether T accepts v_i^* or not. Here, C gets message x_b from the bulletin board immediately after x_b is put on the board. $View_C(X : V_i)$ means C's view through communicating with V_i and getting information from the bulletin board, that* is, $View_C(X : V_i)$ includes published information X, C's coin flips, v_i^*, and the messages that C receives from V_i.

A voting system is receipt-free, if there exists a voter, V_i, such that, for any adversary C, V_i can cast v_i $(v_i \neq v_i^*)$ which is accepted by T, under the condition that $View_C(X : V_i)$ is accepted by C.

Note: In the above-mentioned definition, we assume that the final voting result (total number of votes for each candidate) does not affect the decision of whether C accepts $View_C(X : V_i)$ or not. That is, the total number of votes for v_i^* changes by 1 depending on whether V_i casts v_i^* or v_i $(v_i \neq v_i^*)$. We assume that C is insensitive to such change in the total number of votes. (This assumption is very reasonable, since at least the voting result must be published in any voting system.)

5 Modified voting scheme using untappable channels (Scheme A)

Here, we assume an untappable channel and the parameter registration committee (PRC).

5.1 Untappable channel

Definition 5.1 A physical apparatus is called an "untappable channel" for voter V_i, if only V_i can send out a message, m, to recipient R, and all others can know (information theoretically) nothing about m.

Let $R_1, \ldots, R_N$ be PRC members.

5.2 Procedures

[Authorizing stage]

Public parameters are the same as the original scheme.

V_i randomly generates $\alpha_i \in Z_q$, and splits α_i into N pieces, $\alpha_{i,1} \ldots, \alpha_{i,N}$ such that $\alpha_i = \alpha_{i,1} + \cdots \alpha_{i,N} \bmod q$. V_i then calculates $G_i = g^{\alpha_i} \bmod p$, and $G_{i,j} = g^{\alpha_{i,j}} \bmod p$ $(j = 1, \ldots, N)$.

The other procedure in this stage is the same as the original except

$$x_i = H(m_i \| G_i \| G_{i,1} \| \cdots \| G_{i,N}) t_i^e \bmod n.$$

Therefore, finally V_i gets A's blind signature s_i of $(m_i \| G_i \| G_{i,1} \| \cdots \| G_{i,N})$.

[Voting stage]

V_i sends $(m_i \| G_i \| G_{i,1} \| \cdots \| G_{i,N}, s_i)$ to the bulletin board through an anonymous channel. V_i also sends (v_i, r_i, m_i) to timeliness commission member T through an untappable anonymous channel.

In addition, V_i sends $\alpha_{i,j}$ to R_j $(j = 1, \ldots, N)$ along with G_i through an untappable anonymous channel.

R_j calculates

$$G_{i,j} = g^{\alpha_{i,j}} \bmod p,$$

and sends $G_{i,j}$ along with G_i to the bulletin board.

[Claiming stage]

V_i checks that his/her ballot is listed on the bulletin board (ballot list). If his/her vote is not listed, then V_i claims this by showing $(m_i\|G_i\|G_{i,1}\|\cdots\|G_{i,N}, s_i)$.

In addition, V_i checks that all $G_{i,j}$ $(j = 1,\ldots,N)$ are listed on the board by R_j. If $G_{i,j}$ is not listed, then V_i claims this and sends again $\alpha_{i,j}$ to R_j $(j = 1,\ldots,N)$ along with G_i through an untappable anonymous channel.

[Counting stage]

T (and others) checks whether all $G_{i,j}$ of $(m_i\|G_i\|G_{i,1}\|\cdots\|G_{i,N})$ with s_i are the same as $G_{i,j}$ sent by R_j, and $G_i = \prod_{j=1}^{N} G_{i,j}$ mod p. If this check fails, the corresponding vote v_i is removed from the list of votes.

The other procedure is the same as the original one.

5.3 Proof of receipt-freeness

In this subsection, we prove that the above-mentioned modified scheme satisfies receipt-freeness, if all PRC members are honest.

Theorem 5.2 *Let T follow the protocol. Let σ (T's proof) be the interactive version (i.e., perfect zero-knowledge interactive proof). Assume that untappable channels are available and that all PRC members, R_j $(j = 1,\ldots,N)$ follow the protocol. Then the modified voting scheme A satisfies receipt-freeness.*

Proof. Suppose that all procedures for V_i are done by adversary C, except for the procedure of sending messages to R_j and T through untappable channels. That is, the only role of V_i is sending (v_i, r_i, m_i) to T and $\alpha_{i,j}$ to R_j $(j = 1,\ldots,N)$ through untappable channels.

Such adversary C is universal since if a voting scheme is receipt-free for this type of adversary C, then the voting scheme is also receipt-free for any other type of adversary C^+. This is because: Suppose that for any adversary C of this type, there exists a voter, V_i, such that V_i can cast v_i $(v_i \neq v_i^*)$ accepted by T, under that $View_C(X : V_i)$ is accepted by C. Then for any other type of adversary C^+ with more limited view than C, we can construct voter V_i^+ which follows V_i's strategy and adopts any strategy for the part that C^+ does not execute but C executes in place of V_i. Then for any adversary C^+, there exists a voter, V_i^+, such that V_i^+ can cast v_i $(v_i \neq v_i^*)$ accepted by T, under that $View_{C^+}(X : V_i^+)$ is accepted by C^+.

Here, w.l.o.g., we can assume that C accepts $View_C(X : V_i)$ only if the messages sent out by V_i through untappable channels are compatible with $View_C(X : V_i)$ (more precisely C's view except $G_{i,j}$ sent by R_j $(j = 1,\ldots,N)$). That is, we can assume that C accepts $View_C(X : V_i)$ only if $G_{i,j}$ sent by R_j are exactly the same as $G_{i,j}$ authorized by A's signature in the authorizing stage.

If $G_{i,j}$ $(j = 1,\ldots,N))$ are sent by R_j, then R_j receives $\alpha_{i,j}$ from V_i through an untappable channel, under the condition that R_i follows the protocol. Then, V_i must send out $\alpha_{i,j}$ to R_j, under an untappable channel assumption. This means V_i can calculate $\alpha_i = \alpha_{i,1} + \cdots \alpha_{i,N}$ mod q, and then calculate (v_i, r_i) $(v_i \neq v_i^*)$ such that $m_i = BC(v_i^*, r_i^*) = BC(v_i, r_i)$ by using α_i with $v_i + \alpha_i r_i \equiv v_i^* + \alpha_i r_i^* \pmod{q}$. Therefore, if V_i can send out messages to R_j which are compatible with $View_C(X : V_i)$, then V_i can calculate (v_i, r_i) $(v_i \neq v_i^*)$ such that $m_i = BC(v_i^*, r_i^*) = BC(v_i, r_i)$.

Let V_i^* be V_i who follows C's coercion (i.e., V_i^* casts v_i^* to T). Let V_i cast v_i to T $(v_i \neq v_i^*)$ under the condition that V_i sends out messages to R_j which are compatible with $View_C(X : V_i)$. W.l.o.g., we can suppose that C accepts $View_C(X : V_i^*)$.

Now we assume that C does not accept $View_C(X : V_i)$. The only difference between $View_C(X : V_i^*)$ and $View_C(X : V_i)$ is the voting result and T's proof (say $(Res_{V_i}, \sigma_{V_i})$ with V_i and $(Res_{V_i^*}, \sigma_{V_i^*})$ with V_i^*). This means C can distinguish between the $(Res_{V_i}, \sigma_{V_i})$ and $(Res_{V_i^*}, \sigma_{V_i^*})$. Since $\sigma_{V_i})$ and $\sigma_{V_i^*})$ are perfectly indistinguishable, C should distinguish Res_{V_i} and $Res_{V_i^*}$. This contradicts the assumption described in the definition of receipt-freeness.

Hence C accepts $View_C(X : V_i)$ when V_i casts v_i $(v_i \neq v_i^*)$ to T who accepts v_i.

6 Modified voting scheme using untappable channels (Scheme B)

In the above-mentioned modified voting scheme, α_i is simply split into N pieces. Therefore, if even one PRC member, R_j, does not follow the protocol, then the receipt-freeness cannot be guaranteed.

In this section, we propose a scheme proof against some faulty PRC members. The scheme uses Feldman-Pedersen's VSS directly [Fel87, Ped91a].

6.1 Procedures

Almost all procedures are similar to the previous scheme except the following part:

Let $K \leq N$. V_i randomly generates $\alpha_i \in Z_q$, and $a_k \in Z_q$ $(k = 1, \ldots, K-1)$. Let $f(x) = \alpha_i + a_1 x + \cdots + a_{K-1} x^{K-1}$, and $\alpha_{i,j} = f(j) \bmod q$ $(j = 1, \ldots, N)$. V_i then calculates $G_i = g^{\alpha_i} \bmod p$, $G_{i,j} = g^{\alpha_{i,j}} \bmod p$ $(j = 1, \ldots, N)$, $F_{i,k} = g^{a_k} \bmod p$ $(k = 1, \ldots, K-1)$.

In the voting stage, V_i sends $(m_i\|G_i\|G_{i,1}\|\cdots\|G_{i,N}\|F_{i,1}\|\cdots\|F_{i,K-1}, s_i)$ to the bulletin board through an anonymous channel. V_i also sends $\alpha_{i,j}$ to R_j $(j = 1, \ldots, N)$ along with G_i through an untappable anonymous channel. R_j calculates

$$G_{i,j} = g^{\alpha_{i,j}} \bmod p,$$

and sends $G_{i,j}$ along with G_i to the bulletin board.

In the counting stage, T (and others) check whether all $G_{i,j}$ of $(m_i\|G_i\|G_{i,1}\| \cdots \|G_{i,N}\|F_{i,1}\|\cdots\|F_{i,K-1})$ with s_i are the same as the $G_{i,j}$ sent by R_j, and

$$G_{i,j} = G_i \prod_{k=1}^{K-1} F_{i,k}^{j^k} \bmod p.$$

6.2 Receipt-freeness

Theorem 6.1 Let T follow the protocol. Let σ (T's proof) be the interactive version (i.e., perfect zero-knowledge interactive proof). Assume that untappable channels are available and that at least K PRC members among $\{R_1, \ldots, R_N\}$ follow the protocol. Then the modified voting scheme B satisfies receipt-freeness.

The proof uses the known results on Feldman-Pedersen's VSS and the same techniques used in the proof of the previous theorem.

This scheme can be extended to the unconditionally secure (for $G_{i,j}$ and $F_{i,k}$) version based on the unconditionally secure VSS by Pedersen [Ped91b].

7 Modified voting scheme using voting booths (Scheme C)

In this section, we assume a voting booth, which is a stronger physical assumption than an untappable channel, but we do not need the help of the voting commission.

7.1 Voting booth

Definition 7.1 *A physical apparatus is called "voting booth" for voter V_i, if only V_i can interactively communicate with another party R through the booth, and all others can know (information theoretically) nothing about the communication.*

We also require an additional property, anonymity for the voting booth, i.e., R does not know who V_i is.

7.2 Procedures

[Authorizing stage]

All procedures in this stage are the same as in the original.

[Voting stage]

The procedures in this stage are the same as in the original, except for an additional procedure as follows:

V_i proves to T through an anonymous voting booth that V_i knows α_i in a zero-knowledge manner [TW87] (or with a more efficient protocol such as [Sch91] in practice). If T accepts V_i's proof, then T accepts his vote, $(m_i\|G_i, s_i)$, under the condition that the vote is also valid.

[Claiming stage]

The procedures in this stage are the same as in the original, except for the claiming procedure as follows:

If V_i's vote is not listed on the bulletin board, V_i claims this by showing $(m_i\|G_i, s_i)$ and proving to T through the anonymous voting booth that V_i knows α_i in a zero-knowledge manner [TW87].

[Counting stage]

The procedure in this stage is the same as the original.

7.3 Receipt-freeness

Theorem 7.2 *Let T follow the protocol. Let σ (T's proof) be the interactive version (i.e., perfect zero-knowledge interactive proof). Assume that voting booths are available. Then the modified voting scheme C satisfies receipt-freeness.*

8 Remarks on the security of multiple timeliness commission members

This section shows some remarks for the case of using multiple timeliness commission members (Section 5 in [Oka96]):

- Each T_l ($l = 1, 2, \ldots, L$) sends each v_{il} to their private board (for T_l) and calculate $v_i = v_{i1} + \cdots + v_{iL} \bmod q$. In this stage, we assume that T_l sends $BC(v_{il})$ and then reveals v_{il} after all T_l sends $BC(v_{il})$. Here, BC is a standard bit-commitment in which only a unique value can be revealed after fixing $BC(v_{il})$.
- When the voting is tally, v_i should be multiple bits long with redundant bits for error detection. In other words, one bit ballot should be coded by an error correcting or detecting code.
- The random permutatios π and δ should be split to L timeliness commission members, T_l. That is, each T_l generates random permutatios π_l and δ_l individually, and $\pi = \pi_1 \circ \cdots \circ \pi_L$ and $\delta = \delta_1 \circ \cdots \circ \delta_L$. The basic idea (one-round version) is as follows:

  o V_i splits $v_i = v_{i1} + \cdots + v_{iL} \bmod q$, $r_i = r_{i1} + \cdots + r_{iL} \bmod q$, and votes

  $$E_1(m_i \| (v_{i1}, r_{i1}) \| E_2((v_{i2}, r_{i2}) \| E_3(\cdots E_L(v_{i2}, r_{i2}) \cdots)$$

  to T_1.
  o These messages are decrypted sequentially by T_1 through T_L in a Mixnet manner [Cha81]. The permutations in the Mixnet-like tramsmission from T_l to T_{l+1} corresponds to π_l ($l = 1, \cdots, L$).
  More precisely, in the first round, T_l ($l = 1, \cdots, L-1$) descrypts $E_l((v_{il}, r_{il}) \| E_{l+1}(\cdots)$ and obtains (v_{il}, r_{il}) and calculates $B_{i,l} = B_{i,l-1} H^{v_{il}} \bmod P$. T_l sends $B_{i,l}$ and $E_{l+1}(\cdots)$ to T_{l+1} in a permutated order with π_l. (Here, $B_{i,0} = 1$, P is a prime, and order of H over $\mathrm{GF}(P)^*$ is Q.) Finally, T_L sends $B_{i,L}$ to their private board (T_{L+1}).
  In the second rounds, T_l sends $D_{i,l-1} + v_{il} \bmod Q$ to T_{l+1} in the same order as above, and T_{l+1} checks whether $B_{i,l} = H^{D_{i,l}} \bmod P$ holds. Finally, T_L sends $D_{i,L} = v_i$ to the board (T_{L+1}). Here, $\{v_i\}$ is published in a permutated order with π from the order of $\{(m_i, G_i)\}$. (Each T_i checks whether $B_{i,L} = H^{v_i} \bmod P$ holds.)
  In the third round, the message flow direction is reverse (from T_L to T_1). T_l randomly selects s_{il} and u_{il}, calculates $Z_{i,j,l} = Z_{i,j,l+1} G_j^{s_{il}} h^{u_{il}} \bmod p$, and sends $\{(Z_{i,1,l}, \cdots, Z_{i,5,l})\}$ to T_{l-1} in the reverse order to above (i.e., π_l^{-1}). Here, $Z_{i,j,L+1} = 1$. Finally, T_1 calculates $Z_i = m_i Z_{i,i,1} \bmod p$ and publishes $(Z_1, \cdots, Z_5)$. Here, $Z_{i,i,1} = G_i^{s_i} h^{u_i} \bmod p$, $s_i = s_{i1} + \cdots + s_{iL} \bmod q$, $u_i = u_{i1} + \cdots + u_{iL} \bmod q$.

o Next, the message flow direction is from T_L to T_1, and T_l generates another random permutation δ_l $(l = L, \cdots, 1)$. T_l randomly generates t_{il} and $W_{i,l} = W_{i,l+1} h^{t_{il}} \bmod p$, and sends it to T_{l-1} in a permutated order with δ_l. Here $W_{i,L+1} = g^{v_i} \bmod p$. Finally, T_1 sends $W_{i,1} = W_i$ to the board (T_0). Here, $W_i = g^{v_i} h^{t_i}$ $t_i = t_{i1} + \cdots + t_{iL} \bmod q$.

o Then the interactive proof σ is generated by the collaboration of T_1 through T_L.

If $e = 0$, then T_1 through T_L collaboratedly calculate $s_i = s_{i1} + \cdots + s_{iL} \bmod q$, $u_i = u_{i1} + \cdots + u_{iL} \bmod q$, $t_i = t_{i1} + \cdots + t_{iL} \bmod q$, $\delta = \delta_L \circ \cdots \circ \delta_1$. (e.g., T_l sends $S_{i,l} = S_{i,l+1} + s_{il} \bmod q$ to T_{l-1}, where $s_i = S_{i,1}$, and $S_{i,L+1} = 0$.)

If $e = 1$, then T_1 through T_L collaboratedly calculate $\rho = \rho_L \circ \cdots \circ \rho_1$ (where $\rho_l = \delta_l^{-1} \circ \pi_l^{-1}$), $x_i = \sum_{l=1}^{L} (r_{il} + s_{il}) \bmod q$, $y_i = \sum_{l=1}^{L} (u_{il} - t_{\rho(i)l}) \bmod q$.

Acknowledgments

The author would like to thank Kazue Sako and Markus Michels for pointing out the security flaws of the previous scheme and for invaluable discussions on the receipt-free voting schemes.

References

BGW88. M. Ben-Or, S. Goldwasser, and A. Wigderson, "Completeness Theorems for Non-Cryptographic Fault-Tolerant Distributed Computation", Proc. of STOC'88, pp.1–10 (1988).

BT94. J. Benaloh and D. Tuinstra, "Receipt-Free Secret-Ballot Elections", Proc. of STOC'94, pp.544–553 (1994).

BY86. J. Benaloh and M. Yung, "Distributing the Power of a Government to Enhance the Privacy of Votes", Proc. of PODC'86, pp.52–62 (1986).

Cha81. D. Chaum, "Untraceable Electronic Mail, Return Addresses, and Digital Pseudonyms", Communications of the ACM, Vol.24, No.2, pp.84–88 (1981).

Cha85. D. Chaum, "Security without Identification: Transaction systems to Make Big Brother Obsolete", Communications of the ACM, Vol.28, No.10, pp.1030–1044 (1985).

Cha88. D. Chaum, "Elections with Unconditionally-Secret Ballots and Disruption Equivalent to Breaking RSA", Proceedings of Eurocrypt'88, LNCS 330, Springer–Verlag, pp.177–182 (1988).

CCD88. D. Chaum, C. Crépeau, and I. Damgård, "Multiparty Unconditionally Secure Protocols", Proc. of STOC'88, pp.11–19 (1988).

CF85. J. Cohen and M. Fisher, "A Robust and Verifiable Cryptographically Secure Election Scheme", Proc. of FOCS, pp.372–382 (1985).

CFSY96. R. Cramer, M. Franklin, B. Schoenmakers, and M. Yung, "Multi-Authority Secret-Ballot Elections with Linear Work", Proc. of Eurocrypt'96, LNCS 1070, Springer–Verlag, pp.72–82 (1996).

Fel87. P. Feldman, "A Practical Scheme for Non-interactive Verifiable Secret Sharing", Proc. of FOCS, pp. 427–437 (1987).

FFS88. Feige, U., Fiat, A. and Shamir, A.: Zero-Knowledge Proofs of Identity, Journal of CRYPTOLOGY, Vol. 1, Number 2 pp.77–94(1988)

FOO92. A. Fujioka, T. Okamoto, and K. Ohta, "A Practical Secret Voting Scheme for Large Scale Elections", Proc. of Auscrypt '92, LNCS, Springer–Verlag, pp. 244–251 (1992).

GMW87. O. Goldreich, S. Micali, and A. Wigderson, "How to Play Any Mental Game, or a Completeness Theorem for Protocols with Honest Majority", Proc. of STOC, pp.218–229 (1987).

Ive92. K. R. Iversen, "A Cryptographic Scheme for Computerized General Elections", Proc. of Crypto '91, LNCS 576, Springer–Verlag, pp.405–419 (1992).

JSI96. M. Jakobsson, K. Sako, and R. Impagliazzo, "Designated Verifier Proofs and Their Applications", Proc. of Eurocrypt '96, LNCS 1070, Springer–Verlag, pp.143–154 (1996).

Mic97. M. Michels, "Comments on a receipt-free voting scheme", manuscript (Jan. 1997).

Oka96. T. Okamoto, "An Electronic Voting Scheme", Proc. of IFIP'96, Advanced IT Tools, Chapman & Hall, pp.21–30 (1996).

TW87. Tompa, M. and Woll, H.: Random SelfReducibility and Zero Knowledge Interactive Proofs of Possession of Information, Proc. of FOCS'87, pp.472-482 (1987).

Ped91a. Pedersen, T. P., "Distributed Provers with Applications to Undeniable Signatures", Proceedings of Eurocrypt 91 (1992).

Ped91b. Pedersen, T. P., "Non-Interactive and Information-Theoretic Secure Verifiable Secret Sharing", Proceedings of Crypto 91, pp. 129–140 (1992).

Sch91. Schnorr, C.P., "Efficient Signature Generation by Smart Cards", Journal of Cryptology, Vol. 4, No. 3, pp.161-174 (1991).

SK94. K. Sako, and J. Kilian, "Secure Voting Using Partially Compatible Homomorphisms", Proc. of Crypto'94, LNCS 839, Springer–Verlag, pp.411–424 (1994)

SK95. K. Sako, and J. Kilian, "Receipt-Free Mix-type Voting Scheme", Proc. of Eurocrypt'95, LNCS 921, Springer–Verlag, pp.393–403 (1995)

Flexible Internet Secure Transactions Based on Collaborative Domains

Eduardo Solana and Jürgen Harms

Université de Genève - Centre Universitaire d'Informatique
24, rue Général-Dufour
CH-1211 Genève 4
{Eduardo.Solana,Juegen.Harms}@cui.unige.ch

Abstract. The absence of manageable global key distribution schemes is seriously hindering the deployment of basic security services in the Internet. The emergence of cryptosystems based on public key technology has represented a significant improvement in this direction by removing the need of a mutual agreement on the encryption key. However, the certificate structures that bind a user to his public key are difficult to deploy especially in inter-domain environments. As a consequence, although the need for security services like encryption or authentication is becoming crucial, most Internet transactions currently take place without the use of any of these services. This paper proposes a novel approach for simplifying key manageability relying on the notion of security domains. The fundamental idea relies on the fact that key management and thus security services are easier to achieve inside a well confined domain. Consequently, large scale security might be seen as a combination of intra-domain security and a secure framework for transactions between domains. In other words, user keys are managed internally and only domain keys need to be handled globally. We present the cryptographic schemes needed to achieve confidentiality and authentication based on the collaboration of security domains.

1 Introduction

The need for security in large scale networks such as the Internet is constantly increasing. The enormous growth of the user community and especially the opening to other sectors than pure academic ones has brought a significant amount of potential customers to the electronic market. Since this effect was not foreseen by the conceivers of the Internet, the existing facilities do not fulfill the security requirements of the new transactions (commerce on the WWW, secure E-mail, digital cash, etc.).

Consequently, a strong research and commercial activity has been pursued in the last few years to provide Internet transactions with adequate security features. These activities cover the whole spectrum of the TCP/IP protocol stack and go from the design of a security architecture at the network and transport level (IPnG [1], SSL [2]) to solutions conceived for specific applications

(S-HTTP [3] for the World Wide Web, PEM [4] for electronic mail, etc.). Unfortunately, none of these efforts has resulted in an approved, universally accepted solution. Therefore, security in Internet transactions has been developed quite anarchically and suffers from an "All or Nothing" effect. This means, organizations concerned by security issues conceive strong internal security policies and interact with the Internet through very restrictive firewalls or by means of well-protected *Virtual Private Networks* (*VPN*). Most of the remaining organizations suffer from a complete lack of security in their internal and external transactions (passwords transmitted in the clear, no provision for confidentiality/authentication, etc.). In our opinion, this situation is caused by the following factors:

- **The pioneering Internet spirit:** Although in the last few years the Internet has gained enormous relevance in commercial environments, the open spirit (unrestricted "global connectivity") of the Internet remains alive especially in the academic community. At least so far, this attitude has delayed the design of adequate security solutions.

- **The export and use restrictions of cryptography:** Since cryptographic technology is a fundamental building block for security services, the restrictions that rule the export and use of cryptographic material in certain countries constitute a serious impediment to the world-wide utilization of software containing security components.

- **The lack of a world-wide certificate structure:** Although the emergence of asymmetric cryptographic techniques has suppressed the need for a two-party a priori agreement on the encryption key, the problem of securely binding public keys to entities remains. The undertaken efforts addressing this problem (see next section) have failed to provide a world-wide adapted solution for this issue.

This paper does not address the first two items that should be approached in political rather than scientific terms; the aim of our work consists in analyzing the third point and proposing solutions that simplify the design and management of complementary key structures. This paper proposes the definition and collaboration of *security domains* to improve both the flexibility and the management of these structures. As a consequence, the design and implementation of security policies and, in particular, the deployment of secure transactions in the Internet can be notably simplified.

1.1 Current approaches for providing security

In the field of cryptography, the extensive amount of research pursued has resulted in a very vast choice of algorithms and tested implementations. Apart from isolated remaining struggles - concerning the potential weaknesses of specific algorithms or the reasonable minimum key size - a large consensus has been reached to designate the cryptographic components for securing Internet transactions. This recommendation consists in using (1) fast symmetric algorithms (IDEA [5], DES [6], RC5 [7], etc.) together with randomly generated session keys for encrypting the core of a transaction and (2) asymmetric key technology (RSA [8], Diffie-Hellman [9], etc.) as a support for encrypting session keys that are transmitted within the transaction.

However, the design, implementation and management of a certificate structure binding end-users to their public keys constitute the capital issues for deploying security in a large scale network as the Internet. This point that remains a subject of strong controversy, has resulted in the emergence of different trends:

- The ISO Authentication Framework - and in particular the X.509 protocol [10] - proposes a standardized method to store certificates in a world-wide distributed database which associates end-users to their public keys. The accuracy of this binding is guaranteed by a so-called Certification Authority (CA). The CAs constitute a hierarchical structure where end-users are located at the leaves of the tree. Users obtain the certificates of their peers by establishing a certification path (a chain of signatures) starting at a commonly trusted CA and ending at the peer user entry. This architecture has been selected as key management support for the Privacy Enhanced Mail (PEM [4]), *MIME Object Security Services* (MOSS [11]) and *Secure MIME* (S/MIME [12]). All three aim at providing security to E-mail transactions. The X.509 architecture has also been adopted for other applications such as the *Secure Socket Layer* (SSL [2]) developed by Netscape to secure higher levels of the TCP/IP stack (and notably the WWW transactions). Although new X.509 developments [13] bring forward a higher flexibility (notably allowing for cross certification between CAs), existing implementations remain quite scarce. In our opinion, this is mainly due to the inherent complexity of its certification structure and to the practical difficulty to map existent CAs into a true hierarchical structure.

- Pretty Good Privacy (PGP [14] and PGP/MIME [15]) aims also at providing security enhancements to Internet E-mail. For certificate issues, PGP relies on a transitive chain of trust between users (web of trust) and has gained significant popularity in the Internet community. This certificate structure offers considerable functionality with relatively low administrative overhead to restricted groups of E-mail users. Unfortunately, the PGP certificate structure and its principle of user trust are hardly scalable to the real needs of the Internet because of the vulnerability of the transitive chain when used in loosely connected environments.

- The IETF is currently working in an extension of the *Domain Name System* (DNS) featuring the inclusion of cryptographic parameters (public keys and signatures) as well as reinforced mechanisms to secure naming queries (DNSsec [16]). However, the scaleability of the proposed solution to hold user keys and the definition of a global policy under which these keys will be signed remain uncertain issues [17].

– In the absence of a global solution, another attempt to solve the issue of key management consists in exchanging public keys *manually*. Partners verify the correctness of the key by interchanging the results of a one-way computation (fingerprint). This method is also suitable for applications with very strong security requirements for which the above mechanisms are not secure enough, especially due to the problems of time validity and revocation. A recent paper by Carl M. Ellison [18] proposes a set of cryptographic exchanges allowing to establish identity in the absence of Certification Authorities. Although, these methods answer the needs of specific groups of users, they, obviously, do not fulfil the global requirements of the Internet community.

In a recent work (Mail Ubiquitous Security Extensions - MUSE [19]), Donald E. Eastlake proposes mail relay level and gateway level encryption services to facilitate deployment of electronic mail security. Although the motivations of using coarser granularity are very similar to those presented here, the MUSE approach is *host-based* rather than *domain-based*. As a consequence, there is no support for provision of end-to-end security through the combination of intra and inter-domain security as shown in this paper.

In summary, after several years of efforts, the provision of security services is being seriously retarded by the lack of an adequate key management architecture. Although the inherent complexity of designing a global certificate structure has probably retarded the initial deployment of adapted solutions, we believe that the choice of the user as the finest granularity in a world-wide certificate structure constitutes the most important obstacle for enabling security features to a significant part of the Internet community. And even if an initial deployment of these services were possible, key issues of public key technology such as time validity and certificate revocation would raise unwieldy management problems [20]. Furthermore, this problem will be certainly magnified by the impressive growth rate of the Internet population.

1.2   Domain interventions

The notion of domain appears in many different flavors in the context of the Internet. For instance, at the IP level, the grouping aspect inherent to the domain concept appears in the division of Internet addresses into network, subnetwork and host. This division clearly improves the efficiency of the routing process and allows to specify group-based access control permissions at network level. The hierarchical structure of E-mail addresses (in both SMTP [21] and X.400 [22] protocols) permits to stipulate domain-based routing criteria. Furthermore, the notion of domain appears when a set of equipments or a private network need to be protected behind *firewalls* ([23],[24]) or as a means to facilitate the deployment of access control policies ([25],[26],[27]).

The architectural and naming aspects of a domain inside an organization have been extensively developed by M. Sloman [28]. Consequently, in our work, we assume that the criteria to compose domains (for instance, according to security or administrative needs) are already established and that a naming convention for domains has been defined. We will concentrate exclusively on the impact that the domain collaboration might have in the provision of flexible and manageable mechanisms to secure Internet transactions.

This paper is structured in two major parts. The first one describes the benefits of introducing domains for the provision of security services by comparing them to their user-based counterparts; the second part focuses on the architectural needs of the collaborative domains and explains how the elements of this architecture interact in order to deliver security services to Internet transactions.

2 Domain-based security services

2.1 Domain-based versus User-based: administrative issues

User-based security is appropriate for specific applications, notably those with very strong security requirements. However, as mentioned in the previous section, in large scale inter-networks, a world-wide user certificate infrastructure results in serious implementation and management problems. While relying on the interactions of domains for the achievement of secure transactions, we suggest to modify this certificate structure to handle at its lowest level *security domains* instead of end-users. In other words, we propose a world-wide certificate structure for security domains together with a locally managed, intra-domain architecture containing cryptographic parameters for domain members. From an administrative point of view, this approach presents significant advantages:

- At a global level, the *domain granularity* simplifies both the manageability and the implementation of the global certificate structure. The use of existing and operational naming infrastructures (see previous section) becomes feasible. The coarser granularity *simplifies the management of Certificate Revocation Lists (CRLs)* since only revocation of domain keys needs to be globally handled. As a result, the process of identifying compromised keys is sensibly optimized when compared to the user-based approach where very long CRLs corresponding to a whole CA need to be examined.

- The implementation of a system holding and distributing cryptographic parameters within a security domain (for instance, a local department inside an organization) does not result in excessive management overhead. Several approaches have been successfully implemented: traditional *user/password* mechanisms, public/private keys, Kerberos [25], etc. Even the implementation of an X.509 certificate structure becomes feasible from a management point of view in an intra-domain environment. Furthermore, since end-users are exclusively authenticated at the local level, reuse of existing intra-domain key management solutions becomes possible.

2.2 Providing flexible security services

The security services we aim to provide with the *Collaborative Domains* infrastructure correspond to the basic security services in traditional, user-based security. This is: (a) *confidentiality* and (b) *authentication, integrity* and *non-repudiation*. For simplicity, further on, the term *authentication* will be used to designate the three services of group (b)[1].

In addition to the implementation and administrative advantages, the domain-based security concept brings a higher degree of flexibility by introducing two domain-based alternatives for providing the above mentioned security services with low administrative overhead:

– **Secure, domain-based, inter-domain transactions.** Security services are achieved between domain borders using domain global keys.

– **Secure, domain-based, end-to-end transactions.** The combination of intra-domain (local keys) and inter domain (global keys) encryptions results in *end-to-end* confidentiality and authentication.

– The third option remains the classical, **secure user-based, end-to-end transactions** without domain intervention, where the cryptographic parameters used are, exclusively, those of the principals involved. The finer key granularity of this method makes it especially convenient for applications having very strong security requirements.

The main benefit of the resulting *three-layer* model is that organizations may choose the degree of security they need without having to cope with an unnecessary management and implementation effort. Apart from describing the architectural needs and the involved protocols of the domain-based options, the next section will deal with the level of security and the potential fields of application that should be expected from this alternatives.

3   Collaborative domains architecture

3.1   Basic architectural elements

Consider a generic Internet transaction[2] (such as a WWW query, an E-mail exchange or an rlogin session) taking place between two principals - for instance, a user, an application or a host - for which we wish to provide security services. For simplicity, we divide the principals into two categories: *initiators* (the E-mail sender, the WWW client, or the rlogin user) and *responders* (the E-mail recipient, the WWW server, or the rlogin daemon). Evidently, according to the direction of the dialog, these roles will be alternatively held by both principals. The domains of the initiator and the responder are named source and destination domains respectively (Fig. 1).

---

[1] In public key environments, all three services involve an identical usage of private/public keys at both ends of the transaction.

[2] The term transaction has been preferred to application or service to stress the genericity (levels 4-6 of the TCP/IP stack) of the proposed concepts.

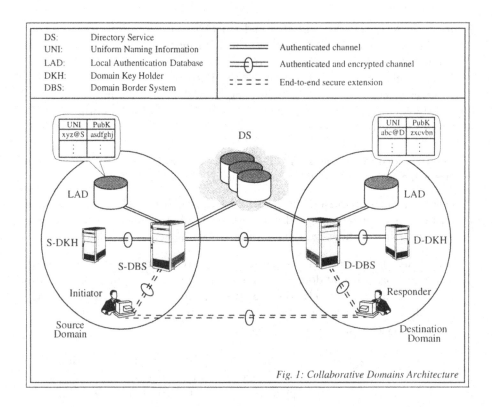

DS:	Directory Service
UNI:	Uniform Naming Information
LAD:	Local Authentication Database
DKH:	Domain Key Holder
DBS:	Domain Border System

═══════ Authenticated channel
══◯══ Authenticated and encrypted channel
⁼ ⁼ ⁼ ⁼ ⁼ End-to-end secure extension

Fig. 1: Collaborative Domains Architecture

A coordinated, global *Directory Service* (DS) holding naming information and especially certificates that securely bind domains to their public keys is also required and constitutes the cryptographic support for inter-domain transactions. As mentioned, existing naming infrastructures (DNS-sec, X.509) might be used for this purpose.

A well defined convention establishing an *Uniform Naming Information* (UNI) is also needed to designate principals and domains globally and unequivocally as, for instance, a common name, an E-mail address, or a network address. Note that this information may also be published in the Directory Service.

The information necessary for the provision of secure end-to-end transactions is kept in the *Local Authentication Database* (LAD) that will contain for each principal (UNI) the local authentication credentials: for this, we propose to use an asymmetric cryptosystem with the principal owning the private key and the LAD holding the corresponding public key[3].

For each security domain, a *Domain Key Holder (DKH) and a Domain Border System* (DBS) are defined[4]. The DKH stores the key-ring of domain public/private key pairs. The DBS is the active player in the domain collaboration and has a dual role:

---

[3] Intra-domain symmetric solutions can be easily adapted to the presented schemes by replacing the LAD by an existing Key Distribution Center (KDC).

[4] Replicating both the DBS and the DKH may be an appropriate policy for performance and reliability issues.

– With respect to inter-domain operations, the DBS provides the required security services (confidentiality, authentication, etc.) between domain borders. For this matter, the DBS needs a secure mechanism (encrypted and authenticated) to obtain private keys from the DKH and public keys of peer-domains from the DS.

– For the provision of end-to-end security services the DBS copes with intra-domain activities such as establishing authenticated or encrypted sessions with the transaction principals. As a consequence, the DBS requires an authenticated access to the LAD in order to verify the identity of principals inside its domain.

Note that the inter-domain operations rely exclusively on cryptographic keys of peer domains provided by a global directory structure whereas intra-domain operations are only ruled by intra domain (locally managed) cryptographic parameters.

The next sub-sections illustrate the use of the proposed approach by describing more in depth the mechanisms involved in the provision of different security services as well as the level of security obtained in each case. Note that only those pieces of information relevant to the encryption process will be mentioned. Other transaction-dependent data records - such as routing information or specific header fields - are omitted to improve clarity of the proposed schemes.

3.2 Confidentiality

Confidentiality prevents information from being disclosed by unauthorized parties. Relying on the described architecture, two different types of confidential transactions can be achieved: *inter-domain* (information is protected exclusively between domain borders) and *end-to-end* (confidentiality is provided in the whole path between principals).

Inter-domain confidentiality. The source DBS acts as *encryptor* for the outgoing transactions, using the public key (obtained from the DS) of the destination domain(s)[5]. Upon data reception, the destination DBS acting as *decryptor* decrypts the transaction with the private key of its domain (obtained from the DKH). According to the volume of information to be transmitted, a DBS may choose to generate a random session key and use it to perform a more efficient, symmetric encryption. In this case, the asymmetric encryption using the public key of the destination domain will exclusively concern a header containing this session key. This header may also contain an identifier for the symmetric algorithm used to encrypt the transaction core in order to ensure compatibility at

---

[5] In case of a multi-cast transaction, a different encryption will be achieved for each destination domain.

both ends. The functionality offered by the DBSs in this scheme is often known as *secure gatewaying*. The main advantage of inter-domain confidentiality lies in the fact that services may be provided *transparently* to the parties involved in the transaction. This is especially convenient for:

– Organizations having well protected private networks which are mostly concerned by securing bulk data exchanges beyond their borders at low management costs.

– Service providers wishing to offer secure transactions as a value added facility independently of the software used at both ends.

End-to-end confidentiality. According to the characteristics of the transaction and the principals involved, we propose two alternatives to achieve end-to-end confidentiality using domain collaborations. The first one, which is a simple extension of the inter-domain mechanism described above, heavily involves the DBS in the encryption and decryption tasks (Fig. 2a):

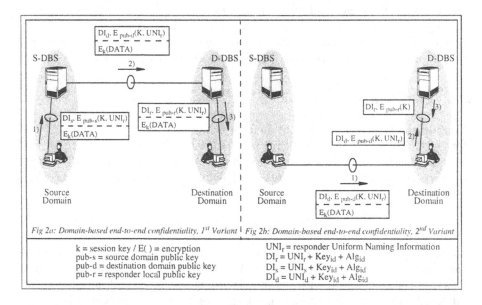

Fig 2a: Domain-based end-to-end confidentiality, 1ˢᵗ Variant | *Fig 2b: Domain-based end-to-end confidentiality, 2ⁿᵈ Variant*

k = session key / E( ) = encryption	UNI = responder Uniform Naming Information
pub-s = source domain public key	$DI_r = UNI_r + Key_{id} + Alg_{id}$
pub-d = destination domain public key	$DI_s = UNI_s + Key_{id} + Alg_{id}$
pub-r = responder local public key	$DI_d = UNI_d + Key_{id} + Alg_{id}$

1. The initiator generates a session key for encrypting transaction contents and creates a header containing the session key and the UNI of the responder. This header is encrypted with the source domain public key and sent to the source DBS together with the transaction core. Additional decryption information (DI) - such as the decryptor UNI and encryption parameters - is sent in the clear to make possible the decryption process.

2. The source DBS decrypts the header using its private key (provided by the DKH), re-encrypts the header with the public key of the destination domain

and sends it to the destination DBS together with the encrypted transaction contents.

3. The destination DBS receives the encrypted packet and extracts the header containing the encrypted session key and the responder UNI. The UNI allows to find the local public key of the responder in the LAD. The session key encrypted with this local public key and the core transaction is forwarded to the responder.

This solution is particularly convenient for principals lacking access to a global DS. It is suitable for domains with strong security policies where the source DBS acts as proxy (unique exit point) and the destination DBS has firewall functionality (unique entry point). Furthermore, the fact that the identity of the responder is encrypted provides an additional level of protection against traffic analysis attacks.

Despite its advantages, this solution introduces significant delay due to the transit of the whole transaction between both DBSs. Furthermore, this approach requires up to three public encryptions/decryptions with different keys which may represent an unacceptable overhead to certain applications. The second method eliminates these drawbacks by having the principals play a major role (Fig. 2b):

1. The initiator generates the same header as in the precedent case (Session Key + responder UNI) and then issues a DS query to obtain the destination domain public key for header encryption. Finally, the whole packet together with the decryption information is submitted directly to the responder.

2. Upon data reception, the responder submits the header to the destination DBS to obtain the session key.

3. The DBS relying on the responder UNI will simply return the session key encrypted with the local public key received from the LAD.

Apart from a lower band-width consumption, an important advantage of this method resides in the fact that the DBS does not have access to the transaction contents.

W. Ford and M.J. Wiener in [29] propose a similar mechanism restricted to E-mail transactions encryption with the aim of providing more flexible access control and object-based protection.

3.3 Authentication

Authentication provides a *proof of identity* of the entity originating the transaction. Two different kinds of authentication are possible: inter-domain where only the originating domain is authenticated and end-to-end where the originating principal is authenticated as being a member of the source domain.

Inter-domain authentication. A DBS acting as *authenticator* should: (a) generate an *integrity code* (IC) by applying a one-way function (*i.e.* MD5) to the whole transaction data (or to a significant part of it); (b) sign the IC with the domain private key obtained through a secure channel from the DKH. The destination DBS acting as *verifier* will validate the authenticated IC using the public key of the source domain obtained from the DS. If the destination DBS has firewall functionality, it may reject transactions coming from unauthenticated domains.

Using an image from the commercial world, this method provides a degree of authentication equivalent to an *enterprise stamp* that would be issued to validate a document. This kind of authentication is valuable for authorizing transactions coming from well-defined domains which might be low-level (a set of equipments) or high-level (a set of applications or users). The described mechanism is also convenient for integrity and non-repudiation provision on a domain basis.

End-to-End authentication. For the provision of end-to-end authentication, we also propose two alternatives. The first one, involving the source DBS is dual to its confidentiality counterpart (see Fig. 3a). Note that upon reception of the authenticated packet, the destination DBS signs it with its private key. As a consequence, the responder needs to trust the destination DBS as a *signature relay* for all the authenticated packets arriving to the domain. Although, this method presents obvious overhead problems, it might be adequate for end systems having limited access to the Internet.

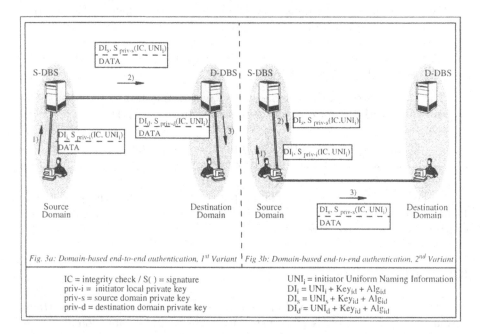

Fig. 3a: Domain-based end-to-end authentication, 1st Variant | *Fig 3b: Domain-based end-to-end authentication, 2nd Variant*

IC = integrity check / S( ) = signature
priv-i = initiator local private key
priv-s = source domain private key
priv-d = destination domain private key

UNI_i = initiator Uniform Naming Information
$DI_i = UNI_i + Key_{id} + Alg_{id}$
$DI_s = UNI_s + Key_{id} + Alg_{id}$
$DI_d = UNI_d + Key_{id} + Alg_{id}$

A more generic and efficient design where principals play the active role is summarized below (See Fig. 3b):

1. The initiator first generates an IC for the transaction contents. Then, it submits a packet containing the IC and its own UNI signed with its local private key to the source DBS. The decryption information (DI) sent in the clear allows to identify the key to be used by the source DBS to verify the signature.

2. Upon verification of the initiator identity (from the LAD), the source DBS sends the initiator the header containing both the IC and the initiator UNI, signed with the source DBS private key.

3. The initiator submits to the responder both this header and the transaction core. The responder verifies the identity (the UNI) of the initiator by decrypting the header with the source domain public key obtained from the DS.

It should be noted that these transactions must be combined with classical techniques as *time-stamps* [30], *one-time nonces* or *mutual handshaking* [31] in order to prevent replay attacks.

The level of authentication obtained by this process depends obviously on how far the source domain can be trusted. Compared to a classical commercial transaction, the degree of authentication obtained is equivalent to an employee signature authenticated by an employer stamp.

4 Conclusions and future work

Most of the present efforts on providing security to Internet transactions have been motivated by the urgent needs of the growing commercial activity in the Internet. Unfortunately, so far, none of these efforts (X.509, PGP) has resulted in a largely deployed solution. This paper argues that an important reason for this situation resides in the choice of *user granularity* as the basic building block of a world-wide certificate structure. Furthermore, even if global deployment were feasible, issues such as *key time validity* and *certificate revocation* would rapidly make this approach impractical: it would present serious management problems and, as a consequence, be ill-suited for a critical purpose such as securing Internet transactions.

The proposal of introducing *domain collaboration* at a global level and as the principal structuring element will represent a notable progress towards providing a significant degree of *flexibility* for the achievement of secure transactions. As shown in this paper, *domain granularity* allows for both *transparent secure gatewaying* between domains and *end-to-end secure transactions*, and prevents the management problems related to a global structure of *user* certificates.

For those applications with particularly strong security requirements, classical user-based security protocols can coexist with the proposed architecture but should, in our opinion, rely on simpler and safer methods (*manually*) for key agreement and certification.

Some issues of the proposed approach need to be further developed:

- Exposure of Domain Border Systems and Domain Key Holders: Centralizing security in a limited number of systems also concentrates security risks. These systems are potential targets for attacks and have to be protected accordingly.

- Cryptanalysis threats: Since all incoming transaction headers arriving to a given domain are encrypted with the same key, the amount of ciphertext (or even of plaintext/ciphertext pairs) available for cryptanalysis can be considerable. Although today's technology is quite resistant to chosen-plaintext attacks, special care must be observed to design correct timing policies for involved keys, in order to preserve the strength of the whole cryptosystem. Obviously, and according to the kind of application to be secured, efficient handling of compromised or revoked keys needs also to be considered.

- Genericity: Since the present work is in an early stage, its application, so far, has only been analysed for some very specific Internet transactions - such as the WWW and the E-mail. The extension of the proposed model to other types of transactions as well as a richer implementation experience of the proposed architectural elements will contribute to refine the design issues described in this paper.

5   References

[1] S. Bellovin, *Security Concerns for IPng*. Request for Comments 1675, August 1994.

[2] *The SSL protocol*, URL http://home.netscape.com/newsref/std/SSL.html.

[3] *Secure HTTP*, URL http://www.eit.com/projects/s-http/.

[4] J.Linn et al. *Privacy Enhancement for Internet Electronic Mail*: Parts I-IV. Request for Comments 1421-1424,February 1992.

[5] X. Lai and J.Massey. *A proposal for a New Block Encryption Standard.* Advantages in Cryptology - EUROCRYPT '90 Proceedings, Springer-Verlag, 1991, pp. 389-404.

[6] Federal Information Processing Standard (FIPS) 46-1. *Data Encryption Standard*,1977.

[7] RSA Data Security, Inc. *The RC5 encryption algorithm*, URL ftp://ftp.rsa .com/pub/rc5/rc5.ps, 1994.

[8] R. Rivest, A. Shamir, L. Adleman. *A Method for obtaining Digital Signatures and Public Key Cryptosystems.* Communications of the ACM, February 1978.

[9] W. Diffie and M. Hellman. *New directions in Cryptography*. IEEE Transactions on Information Theory, November, 1976.

[10] ISO/IEC 9594-8 (1988). CCITT Information Technology - Open Systems Interconnection - *The Directory: Authetication Framework*. Standard X.509, 1988.

[11] S. Crocker, N. Freed, J. Galvin, S. Murphy, *MIME Object Security Services*. Request for Comments 1848. October 1995.

[12] RSA Data Security, Inc. *Secure MIME*. http:// www.rsa.com/rsa/S-MIME/.

[13] R.Housley, S. Farell et al. *Internet Public Key Infrastructure; Parts I,II and III*. Work in progress. Internet Drafts: <draft-ietf-pkix-ipki-part1-03.txt> and <draft-ietf-pkix-ipki3cmp-00.txt>. December 1996.

[14] P.R. Zimmermann. *PGP Source Code and Internals*. Boston: MIT Press 1995.

[15] M.Elkins. *MIME Security with Pretty Good Privacy*. Request for Comments 2015. October 1996.

[16] D. E. Eastlake. *Domain Name System Security Extensions*. Request for Comments 2065. January 1997.

[17] J.M. Galvin. *Public Key Distribution with Secure DNS*. Proceedings of the 6th USENIX Security Symposium. July, 1996.

[18] C.M. Ellison. *Establishing Identity Without Certification Authorities*. Proceedings of the 6th USENIX Security Symposium. July, 1996.

[19] D.E. Eastlake. *Mail Ubiquitous Security Extensions*. Work in progress. Internet Draft: <draft-eastlake-muse-01.txt>. November, 1996.

[20] D. Davis. *Compliance Defects in Public-Key Cryptography*. Proceedings of the 6th USENIX Security Symposium. July, 1996.

[21] D. Crocker, *Standard for the format of ARPA Internet text messages*. Request for Comments 822. August 1982.

[22] ITU and CIITT - *Data Communications Networks Message Handling Systems*. Recommendations X.400-X.420, 1988.

[23] W.R. Cheswick. *The design of a secure internet gateway*. Proceedings of the Summer USENIX Conference, Anaheim, June 1990.

[24] D.B. Chapman. *Network (in)security through IP packet filtering*. Proceedings of the Third USENIX UNIX Security Symposium, pages 63-76, Baltimore, September 1992.

[25] J. Steiner, C.Neuman, J. Schiller, *Kerberos: An authentication for Open Network Systems*, in Proceedings Winter USENIX Conference, 1988.

[26] N. Yialelis and M.Sloman. *A Security Framework Supporting Domain Based Access Control in Distributed Systems* in Proceedings of Network and Distributed System Security Symposium 1996 (NDSS'96).

[27] L. Badger et. al. *Practical Domain and Type Enforcement for UNIX.* Proceedings of the 1995 IEEE Symposium on Security and Privacy. Oakland, California. May, 1995.

[28] M. Sloman. Domains: *A Framework for Structuring Management Policy* in Network and Distributed Systems Management. pp. 433-453. Addison-Wesley, 1994.

[29] W. Ford and M.J. Wiener. *A key Distribution Method for Object-Based Protection.* 2nd ACM Conference on Computer Communications Security. Fairfax, Virginia. 1994.

[30] D.E. Denning and G.M. Sacco. *Time-stamps in Key Distribution Protocols.* Communications of the ACM, Vol.24, n.8, pp. 533-536. August 1981.

[31] R.M Needham and M.D. Schroeder. *Using Encryption for Authentication in Large Networks of Computers.* Communications of the ACM, Vol.21, n.12, pp. 993-999. December 1978.

How to Build Evidence in a Public-Key Infrastructure for Multi-domain Environments

Bruno Crispo * **

Abstract. We discuss here some of the issues that must be considered to build evidence in an appropriate way in a public-key infrastructure (PKI). Despite the fact that one of the most recurrent motivation by papers advocating the necessity of a PKI, is to support electronic commerce, all the new proposals of PKIs do not define any procedure to specify which evidence must be collected and in which form, when users carry out a commercial transaction.
We think that this is an important issue that requires more attention especially if Internet will succeed to became a marketplace as many people hope. In the conventional world, evidence plays a very important role in any dispute resolution that can occur.
Besides all the services and applications that we can provide to users to facilitate them to buy through the use of PCs, we have to provide them by a sufficiently well founded guarantee that they will be safeguarded against frauds attempts and malicious behaviours.
In this paper, we describe which facilities a PKI must provide in order to build the evidence needed to get that guarantee.

1 Introduction

There are a lot of critical issues worth to be considered and studied, in designing a global PKI, our paper however, focuses mainly on the technical requirements needed to define and associate liability among the entities of the system.
We are particularly interested to design a PKI resilients to manipulations by misbehaving or incompetent entities of the system. Our model do not enforce any particular trust relationships between domains or between entities inside a domain, offering the possibilities of building a PKI based on any trust model required by the particular application environment using that PKI.
Our interest in evidence is motivated mainly by two factors: first dispute resolution is one of the most expensive procedure to go through for any company. Any mechanism that will succeed to keep as low as possible the number of cases where an independent arbitrator must be involved in a dispute resolution will reduce sensibly costs. Second, customers need to be protected by possible abuses from authorities as well as fraud attempts from other customers.

---

* Cambridge University Computer Laboratory Pembroke Street, Cambridge CB2 3QG e-mail Bruno.Crispo@cl.cam.ac.uk

** Universita di Torino - Dipartimento di Scienze dell'Informazione e-mail Bruno.Crispo@di.unito.it

Furthermore we think that these capabilities will make easier to design and provide new services required by PKI's users as notarization and time-stamping [24] services. The design of new metrics of authentication [14, 15, 25] can also be helped by having cryptographically strong evidence kept by each entities of the system, especially to reduce their sensitivity to misbehaviour of entities [15]. We do not think that all the PKIs need to provide the facilities we will describe, but for the PKIs used to support authentication in financial transactions this is highly advisable.

We have already presented in [omitted] a certification scheme that, differently from the existing ones, does not rely upon the trust of the certification authority (CA) - in the sense that the CA is not supposed to be unverifyble trusted - and the scheme provides facilities to verify the trustworthiness of the authorities at any time and to help to resolve disputes, when they arise for non repudiation purpose.

Our solution was proposed for the case of a single domain only and we describe briefly in Section 2, the design principles we chose.

In Section 3 we describe why a PKI should be global and in Section 4, we present the extension of our work to multiple domains. We describe how our system will scale up and how it allows, users of different and mutually untrusted domains to authenticate each others relying as much as possible only upon their own local trusted domain. We introduce a new component: the *Security Embassy* and describe how to use it in a new cross-certification scheme to collect and build proper evidence in order to trace entities' behaviour.

We conclude in Section 5.

2 Untrusted Certification Authority for a Single Domain

The main goal of a public-key infrastructure is to solve the authentication problem. Using public-key cryptography, this means, solving the problem of distribute in a safe and verifyble manner the public key of the parties involved in the communications. By means of electronic key certificate (EKC) this can be achieved without involving, in any transaction, a single distribution centre [20].

The use of EKCs introduces the need of having authorities in order to register, certify and eventually process the request of revoking the binding between principals and public keys. All the PKIs proposed until now, as PGP [23], X.509 [13, 12], IPKI [16, 5, 7, 21], [19] and SET [26] assume that these authorities are trusted, in the sense that they can violate the security policy of users without leaving evidence of their misbehaviour. Moreover they let the possibilities to users to falsely accuse of misbehaviour the authorities that with the existing PKIs are not provided by any method to defend themselves if this is going to happen.

We show that for the purpose of key management, trusted authorities are an unnecessary technical constraint that can be removed enhancing consequently the security and the acceptability of a PKI, especially for commercial applications.

Besides this is necessary to guarantee non-repudiation in the key management service. In many commercial applications the resolution of disputes is one of the most critical problem, then it is advisable to provide a PKI with the necessary facilities to build evidence to make possible to solve disputes as easily as possible. Even if the procedure to interpret this evidence it is not only a technical issue but depend also by the laws enforced, it is our task to provide methods to build and keep the evidence in proper manner. Technical features could then help judges in their work choosing the most effective, reliable and cheaper procedures to solve disputes.

In [3] we proposed a PKI in which every entity of the system In a previous paper [omitted], we proposed a PKI in which the initial assumption was that every entity of the system is untrusted.

Our proposal uses two basic principles: *separation of duties* and *auditing*.

The concept of separation of duties is used to separate the registration procedure and the issue of certificates from the revocation service.

The other building blocks of our scheme are the log audit trails.

Following we give an overview of that system.

2.1 Separation of Duties

Our scheme for a single domain is shown in Figure 1:

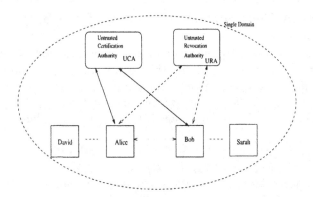

Figure 1. Structure in a Single Domain

We separate the functions of certification and revocation. They are two distinct services with different security requirements and no necessity to be implemented by the same authority. Our system relies upon a server called an *Untrusted Certification Authority* (UCA) for the certification service and an on-line server called an *Untrusted Revocation Authority* (URA) for the revocation service.

The certification facility allow users to be registered in a domain and to have their own EKC issued. The revocation facility allows users to revoke their own public key in case of compromise or loss of the correspondent private key, or in case of change of information used to identify the user, contained in his or her certification request sent to the UCA when he or she asked for the certificate. The URA gets her EKC from her local UCA as any other users of the domain.

Separating the two tasks is necessary in order to enforce the safeguard described above. If the certification and revocation services were to be combined then it is relatively easy to persuade somebody to act upon a fraudulent certificate. In particular by revoking a certificate and then issuing a replacement these services can make it appear that an user has a new key [omitted], if the user subsequently loses his private key, with the current proposals he cannot prove to an arbitrator the authority's misbehaviour.

We do not solve completely the problem but we make more difficult to cheat. Our separation of duty rule means that such an attack would require the collusion of the issuing and revocation authorities.

The manager in charge of the security of one of these services must not be responsible for the security of the other.

2.2 Auditing

As we said all the entities of our PKI are untrusted. To achieve this goal our scheme requires the participants to keep a log of certain parts of the messages that they receive and in particular all that ones containing security sensitive information. Users should then be able to convince a third party of the misbehaviour of somebody that they have communicated with.

Sometimes clients cannot by themselves validate the behaviour of another component, but by cross verifying audit trails of all the parties involved in a dispute an independent arbitrator, in most cases, will be able to validate that behaviour and determine who is the most likely to be blamed for a misbehaviour.

The combination of the audit trail entries with the use of asymmetric cryptography allows us to resolve most plausible attacks.

It is worth noting that not all the evidence is in electronic form, but there are essential part of it that *must be* held in paper form. An example is the letter of authorisation conventionally signed by the requester of an EKC and countersigned by the UCA herself, that both parties must keep safely as evidence, attesting that the named UCA was authorised by the signer to issue that EKC granting only the rights stated in the letter. This letter prevents the UCA to be able to masquerade as the user herself in a different domain and informs the user about all and the only rights for which she will be liable.

Since the audit trails are critical to this resolution it is important that they are held in a tamper-proof form.

Electronic messages that constitute evidence have to be signed[1] and, where confidentiality is considered necessary, they may also be encrypted. Those with signatures should be countersigned by the recipient so that he or she can check for subsequent tampering. It is advisable for these messages also to be countersigned by a third party as proof of an upper bound on their time of origin. While the recipient could forge the date and time on his own signature he, presumably, cannot do so on the countersignature without collusion. We advise the use of notarization or time stamping services for this purpose.

We have seen for the case of a single domain that it is not only possible, but advisable to build a PKI without trusted entities. In the following sections we describe the case for multiple domains.

3 Scaling

The most critical issue still to be solved by all PKI's proposals is scalability.

The only way to be certain that a public key belongs to a particular person and at the same time to be able to convince a third party about this belongings is for that person to physically hand over a copy of their electronic public key and relative paper evidence ,as we have already specified in [3], to whoever they wish to communicate with. Clearly, however, this method of key certification is of little use if it is only effective when one user already knows another well enough to arrange a meeting to exchange keys. Furthermore, in the majority of cases a physical exchange of public keys is impracticable, especially if an user wishes to communicate with a large number of people, some of whom live thousands of miles away. A PKI then, to be useful must be easily scalable and even become hopefully global if necessary.

A user requiring knowledge of a public key generally needs to obtain and validate an EKC containing the required public key. If the user do not hold an assured copy of the public key of the UCA that signed the EKC then he might need an additional EKC to obtain that key. In general a chain of multiple EKCs may be needed, comprising a certificate of the public key owner signed by one UCA, and zero or more additional EKCs of UCAs signed by other UCAs. Such chain is called *certification path* and it is required because a user initially has a limited number (often one) of assured UCA public keys.

UCAs can be structured mainly in two different topology: *hierarchy* and *network*. The systematic, ordered topology of certification paths that is normally employed is a hierarchy [7, 13, 6, 26], while the more general topology is a network of cross-certified UCAs [17].

- *Hierarchical:* UCAs are arranged hierarchical under a "root" UCA, that issues certificates to subordinate UCAs. These UCAs may issue certificates to UCAs below them in the hierarchy, or to users. In a hierarchical PKI, the public key of the root UCA is known to every user, and any user's EKC

---

[1] Through all the paper when we said 'signed' we mean 'electronically signed' otherwise we specify 'conventionally signed'

may be verified by verifying the certification path of EKCs that leads back
to the root UCA.
- *Network:* Independent UCAs cross certify each other resulting in a general
network of trust relationship among UCAs. A user knows the public key of
a UCA near himself, generally the one that issued his certificate, and verifies
the EKC of other users by verifying a certification path of EKCs that leads
back to that "trusted" UCA.

The need for both topologies, already investigated by [17, 18, 25] and last
year by [19, 10, 9, 8] is motivated by the following considerations:

1. many commercial and business trust relationships are not necessary hierarchical
2. the compromise of the root private key is almost catastrofic in a hierarchical
   structure
3. needs in particular situation of direct cross-certification, that allow also to
   reduce certification path processing load.

We will describe, in this paper, only the case of networks of UCAs, where
different problems respect than the single domain case arise. While the case of
hierarchical structures is just a generalisation of the single domain that can be
solved applying the same principles we mentioned in Section 2.
   We define requirements and procedures to apply a *cascade* cross-certification
scheme to build networks of UCAs combining single and/or hierarchical domains.

4 Untrusted Certification Authorities for Multiple Domains

As we said one of the biggest challenge that a PKI has to face is the ability to
scale up without lacking security features for its users. Even if several papers [4,
1, 17, 18, 11] have been written about the problem of authentication in a multi-
domain environment, almost all of them do not analyse the consequences on
evidence and log audit trails implied by the combination of single disjoint security
domains in a larger infrastructure of domains interacting among them. We
describe which procedures must be followed accordingly with these consequences
to build evidence.

4.1 Security Embassy

Before describing in detail the cross-certification scheme, we need to introduce
the concept of *Security Embassy* (SE), already appeared in [2].
The goal of a Security Embassy is to build electronic evidence on behalf of her
owner in a remote (untrusted) domain, in a form suitable for possible later disputes.
A SE is a device provided with a public and a private key that allows her owner
to perform some of her functions on-line and remotely, in a different domain, in
a secure and irrefutable way.

The key pair will be installed by the SE's owner. Once the pair is installed it cannot be changed without leaving clear evidence that that operation took place. The private key must be stored safely also if the device is in an hostile environment. The SE contains also one or more public keys which can be configured only by her owner at any time.

In order to satisfy these requirements the SE must be a tamper proof device.

The SE can be configured in a way which restricts the owner of the SE to only the functions he needs for his operational role. A restricted shell can be developed which allows the owner to execute only those commands necessary for his task. He will never directly get superuser privileges. All actions which require a special privileges will be performed by setuid-programs. An audit function is performed also by a setuid-program that logs each attempted login to the SE as well as each command typed by the owner. In this way the onwer activity itself can be monitored by an independent party.

Functionally the SE receives in input some signed data, performs a signature verification with the public key configured, than if the check is successful, she signs these data using her private key.

The SE then stores internally a copy of these signed data as part of her owner's audit trail. Both host domain and SE's owner can check remotely the SE's operativity simply executing a challenge/response protocol. The potential problem of subliminal channels [22] due to this protocol is not critical in this context.

The SE's owner that belong to a certain domain he trusts, can insert his SE in an other, untrusted, domain with the guarantee that she cannot be subverted without leaving evidence. The owner can trust her SE because she can build it by herself or she can freely choose the manufacturer, because the specifications can be formal and public, furthermore she is the only one involved in the key pair generation used by the SE to sign.

In the following figure we will show an example of SE belonging to UCA_B and accommodated in the UCA_A domain with the specification of which key each party is supposed to know to be able to operate. We use the notation K_{A+} for A's public key and K_{A-} for A's private key, $\{M\}_{K_{A-}}$ means the message M signed by K_{A-}.

The security of the host domain is not jeopardised by the presence of SEs because there are no shared resources between SEs and the host domain and because interactions between them can be formalised to be secure. The data that an entity, belonging to the host domain, sends as input, through a dedicated line, to an accommodated SE are signed so they cannot be forged by the SE and if the data submitted as input are carefully chosen according with the security policy of the host domain, whatever function the SE performs cannot breach that security policy.

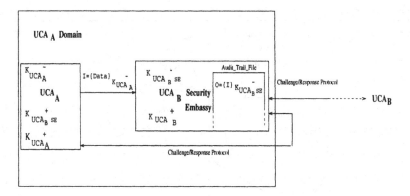

Figure 2. Security Embassy

4.2 Cascade Cross-Certification Scheme

The principle that leads our design is to maintain the trust as much *local* as possible, and delegate it just when it is unavoidable and when this happen anyway providing facilities to be able to verify the (ab)use of this delegation. We define how the trust will be distributed among entities and domains and which protocols to use to implement such distribution.

Let us suppose to have two or more companies, each of them implementing their own UCA. They do not trust completely each other but nevertheless some members of both companies need to communicate, or suppose to have two or more companies wishing their customers to communicate each other for marketing reasons in such a way that the advantages gained by cross-domain transactions could worth the risks associated to possible losses due to the acceptance of liability for other domain's members. Our proposal satisfies these trust models avoiding the unfair solution in which every user is required to go physically to register his public key in every domain he wishes to use.

We propose a system, instead, in which the UCA of the organisation to which he belongs carries out all the procedures to let him to communicate securely with members of other organisations, and at the same time the UCA is liable for the security services she provides.

In order to do so, each UCA has to agree to delegate some of her trust to the remote UCA, with the consequent problem of remote verification.

An UCA has a direct trust relation with users certified locally while she will have an indirect trust relation, by mean of the remote cross-certified UCA, with users certified remotely. The local UCA delegates to the remote one the tasks of registering and issuing EKCs for her remote users. This delegation transforms users of the remote, cross-certified, UCA into local users.

They can use all the security services provided by the local UCA and she will be liable for them, but technically the management of their EKCs is performed to the remote UCA.

The scheme we propose for such situations is the *cascade* cross-certification scheme, shown in Figure 3 for two UCA domains.

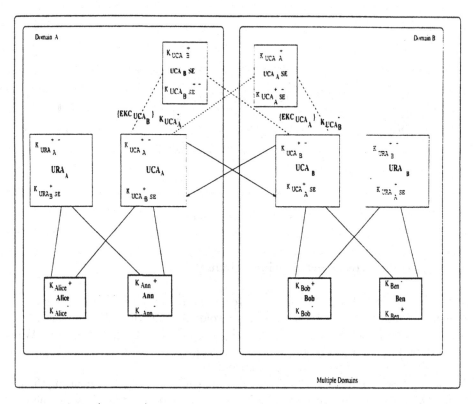

Figure 3. Cascade Cross-Certification Scheme

In case of two UCAs, each of them has to cross certify the other one. The UCAs must set up an initial physical meeting in order to corroborate their own identity reciprocally. In the same meeting they can get in a trusted way the other party's public key.

Besides the public key, they have to exchange a letter of authorisation conventionally signed on site by both UCAs - this contains proof of the value of the key to be cross-certified and the security policy under which the two UCAs agree to cross-certify each other. This paper evidence, one for each UCA, is built and exchanged simultaneously between both authorities during the meeting.

After this initial step UCA_A has issued $\{EKC_{UCA_B}\}_{K_{UCA_A}-}$ while UCA_B has issued $\{EKC_{UCA_A}\}_{K_{UCA_B}-}$.

Where $\{EKC_B\}_{K_A-}$ means the electronic key certificate signed using K_A, binding B with B's public key. It is outside the scope of this paper investigating about the nature of B, if it should be a key, an address or a name.

$\{EKC_{UCA_B}\}_{K_{UCA_A}-}$ together with the paper evidence mentioned above constitute the evidence with which the liability is transferred from UCA_B to UCA_A for what concern her duties as UCA. The UCA_A before accepting the transfer of liability must check that her security policy is compatible with the security policy of the domain she wishes to cross-certify. What is peculiar of our scheme is that this transfer of trust is not accepted unconditionally and moreover it can be verified by UCA_A anytime and using only information belonging to her.

UCA_A in fact will accept liability for UCA_B's member only after she installs her own SE ($UCA_A SE$) into UCA_B domain. The main goal of this SE is to build evidence about the UCA_B activity on behalf of UCA_A in a trusted manner.

When the two UCAs accept to cross-certify each other, implicitly they have to consent to the other UCA to monitor their own activity concerned with the users they share.

In more detail this means that all the messages that UCA_B store in her own audit trail and that are all signed with her private key, must also be sent to the $UCA_A SE$ that notarise and countersigning them, and then she stores them to her internal cryptographically protected, as specified in section 2.2, audit trail.

This protocols allow UCA_A to build on-line evidence about UCA_B activity, after the cross-certification agreement with UCA_B, in a secure manner even if the domain is not completely trusted and remote.

UCA_B in fact, cannot subsequently repudiate messages notarised by $UCA_A SE$ because they were signed by her, before been processed, and for the same reason UCA_A cannot forge them.

Finally UCA_A must be guaranteed about the state of UCA_B when the agreement starts to be valid. UCA_A gets this guarantee, obtaining from UCA_B a signed copy of the user's EKCs issued by her until that date. We do not consider the case of other already existing cross-certification EKCs. This copy is unforgeable because is signed by UCA_B. A good practice for both UCAs is to have this list countersigned by one or more independent parties to prevent the case in which revoking her own key one of the two UCAs could falsely repudiate the accuracy of this copy.

The same protocol is needed by UCA_B to accept liability for UCA_A's users. From users point of view when Alice wishes to communicate with Bob, supposing that they belong to two different and respectively cross-certified domains, she will send the following message:

$$\text{Alice} \rightarrow \text{Bob}: \{message\}_{K_{Alice}-}, \{EKC_{Alice}\}_{K_{UCA_A}-}, \{EKC_{K_{UCA_A}}\}_{K_{UCA_B}-}$$

This chain means that UCA_B says that UCA_A's public key is valid and says Alice's public key is the one contained in the EKC_{Alice}.

When Bob receives this message he can verify $\{EKC_{UCA_A}\}_{K_{UCA_B}-}$, - UCA_A's certificate - using his local UCA's public key, then he gets Alice's public key to validate the origin of the message.

It is called cascade cross-certification because the chain of EKCs that Alice must send to Bob are certified in cascade. The scheme can be easily extended to N UCAs, in that case the chain of EKCs will be certified in cascade N times.

Besides an analogue protocol must be followed for similar reasons by the URAs to build in an appropriate manner their own evidence. In case of URAs they also need to install SEs to monitor the other URA's activity, but differently from the UCAs, they do not need any cross-certification procedure because they rely upon the EKC issued by their respective UCAs. The need of having SE in this case, is due to the delegation of trust that occurs also in case of URAs. The local URA delegates the task of revoking EKCs to the remote URA for her remote users, provided that a cross-certification agreement exists between the respective UCA to which they belong.

In case of URAs, instead of a copy of all the user's EKCs, the parties must exchange initially a signed copy of the current *Certificate Revocation List* (CRL), the list of all the user's EKCs revoked until that time, even if this is empty.

Supposing that Bob has received the message specified above and he wishes to check if Alice's public key is revoked, then the following communication takes place:

Bob $\rightarrow URA_A$: $\{CRLrequest, Bob\}$

$URA_A \rightarrow$ Bob: $\{CRL\}_{K_{URA_A}-}, \{EKC_{URA_A}\}_{K_{UCA_A}-}, \{EKC_{K_{UCA_A}}\}_{K_{UCA_B}-}$

If the above procedure have been applied when a dispute arise all the entities of the system will have, locally, their own technical evidence to face it without need to rely upon somebody else competence.

5 Conclusion

In this paper we described some of the work we are developing to specify requirements and protocols to build electronic and cryptographically strong evidence as well as paper evidence, in a public key infrastructure used as support for commercial transactions. For commercial applications resolution of disputes and non-repudiation are important and critical issues.

We have shown that do not having trusted authorities is a necessary requirement to guarantee a framework to resolve easily disputes.

In a PKI, concentration of trust can potentially turn in ambiguity. Ambiguity does not only make resolution of disputes longer and then more expansive but it provides itself reasons for disputes to arise. We presented briefly our solution in case of a single domain and then we described how to combine single domains. This combination do not enforce any particular trust model.

To avoid recurrent and inconvenient physical meetings, delegation of trust is needed. This delegation arises the problem for the delegator to verify remotely how this delegation is used in order to detect possible misbehaviours by the delegee. We introduced a new component, the Security Embassy to solve this problem efficiently.

Actually we are working to specify more in detail the internal architecture of the SE and eventually to implement it.

6 Acknowledgement:

The author wish to thank Bruce Christianson for his suggestions and help in improving this paper.

He would like also to thank the EPSRC (British Engineering and Physical Sciences Research Council) for supporting his research at the University of Cambridge into Trust in Distributed Systems.

References

1. R. Needham A. Birrell, B. Lampson and M. Schroeder. A Global Authentication Service Without Global Trust. In *Proceedings of the IEEE Conference on Security and Privacy*, 1986.
2. P. Hu B. Christianson and F. Snook. File server architecture for an open distributed document system. In *Communication and Multimedia Security*. Chapman and Hall, 1995.
3. M. Lomas B. Crispo. A Certification Scheme for Electronic Commerce. In *Security Protocol Workshop*, volume LNCS series vol. 1189. Springer-Verlag, 1997.
4. M. Burrows B. Lampson, M. Abadi and E. Wobber. Theory and Practice. *ACM Transaction on Computer Systems*, 10:265–310, November 1992.
5. S. Boeyen, R. Housley, T. Howes, M. Myers, and P. Richard. *Internet Draft: Internet Public Key Infrastructure Part 2: Operational Protocols.* working draft 'in progress' available at:, ftp://ietf.org/internet-drafts/draft-ietf-pkix-ipki2opp-00.txt, March 1997.
6. W. E. Burr. *Public Key Infrastructure - Technical Specifications (version 2.3) Part C - Concept of Operation.* Work 'in progress' available at: http://csrc.nist.gov/pki/twg/conops.ps, November 1996.
7. S. Farrell C. Adams. *Internet Draft: Internet Public Key Infrastructure Part III: Certificate Management Protocols.* working draft 'in progress' available at:, ftp://ietf.org/internet-drafts/draft-ietf-pkix-ipki3cmp-02.txt, June 1997.
8. D.W.Chadwick, A.J. Young, and N.K.Cicovic. Merging and Extending the PGP and PEM Trust Models - The ICE-TEL Trust Model. *IEEE Network*, June/July, 1997.
9. C.M. Ellison. *Internet Draft: SPKI Requirements.* March 97, working draft 'in progress' available at:, ftp://ietf.org/internet-drafts/draft-ietf-spki-cert-req-00.txt, March 1997.
10. C.M. Ellison, B. Frantz, B. Lampson, R. Rivest, B.M. Thomas, and T. Ylonen. *Internet Draft: Simple Public Key Certificate Requirements.* July 97, working draft 'in progress' available at:, ftp://ietf.org/internet-drafts/draft-ietf-spki-cert-structure-02.txt, July 1997.
11. D. Gligor, S Luan, and J.N.Pato. On Inter-Realm Authentication in Large Distributed Systems. In *Proceedings of the IEEE Conference on Security and Privacy*, 1992.
12. Information Technology - Open Systems Interconnection, Geneva. *Draft Technical Corrigendum to Recommendation X.509*, December 1995. The Directory: Authentication Framework.
13. Information Technology - Open Systems Interconnection, Geneva. *Recommendation X.509*, June 1995. The Directory: Authentication Framework.

14. U. Maurer. Modeling a public-key infrastructure. In *ESORICS 96*, Roma, 1996. Springler-Verlag N.Y. LNCS 1146.

15. M.K.Reiter and S.G.Stubblebine. Toward Acceptable Metrics of Authentication. In *Proceedings of the IEEE Conference on Security and Privacy*, 1997.

16. W. Polk R. Housley, W. Ford and D. Solo. *Internet Draft: Internet Public Key Infrastructure Part I: X.509 Certificate and CRL Profile.* working draft 'in progress' available at:, ftp://ietf.org/internet-drafts/draft-ietf-pkix-ipki-part1-04.txt, March 1996.

17. T. Beth R. Yahalom, B. Klein. Trust Relationships in a Secure Systems - A Distributed Authentication Perspective. In *Proceedings of the IEEE Conference on Security and Privacy*, May 1993.

18. T. Beth R. Yahalom, B. Klein. Trust-based navigation in distributed systems. *Computing System 7(1)*, Winter, 1994.

19. B. Lampson R.L. Rivest. SDSI - A Simple Distributed Security Infrastructure. http://theory.lcs.mit.edu/ cis/sdsi.html, April 1996.

20. M.D. Schroeder R.M. Needham. Using Encryption for Authentication in Large Network of Computers. *Communications of ACM*, 21:993–999, 1978.

21. W. Ford S. Chokhani. *Internet Draft: Internet Public Key Infrastructure Part IV: Certificate Policy and Certification Practices Framework.* working draft 'in progress' available at:, ftp://ietf.org/internet-drafts/draft-ietf-pkix-ipki-part4-01.txt, July 1997.

22. Gustav J. Simmons. Subliminal Channel:Past and Present. *Europen Transactions on Telecommunications*, 5:459–473, July-August 1994.

23. W. Stallings. *Protect Your Privacy.* Prentice-Hall, Englewood Cliffs, New Jersey, 1995.

24. S. Stornetta and S. Haber. Secure Digital Names. In *Proceedings of the 4th ACM Conference on Computer and Communication Security*, 1997.

25. B. Klein T. Beth, M. Borcherding. Valutation of trust in open network. In *ESORICS 94*, Brighton, Uk, November 1985. Springler-Verlag N.Y.

26. VISA and MaserCard, http://www.mastercard.com/set or http://www.visa.com/set. *SET: Secure Electronic Transactions* , 1995.

On Signature Schemes with Threshold Verification Detecting Malicious Verifiers

Holger Petersen[1]

Markus Michels

Laboratoire d'informatique
Ecole Normale Supérieure
45, rue d'Ulm
F-75005 Paris, France
hpetersen@geocities.com

Union Bank of Switzerland
Ubilab
Bahnhofsstrasse 45
CH-8021 Zurich, Switzerland
Markus.Michels@ubs.com

Abstract. While in the ordinary digital signature concept one verifier is sufficient to check the validity of a given signature, there are situations in which only t out of a group of n verifiers should be able to verify the signatures. These verifiers can either be anonymous, non-anonymous or convertible non-anonymous. So far, only schemes for the anonymous shared verification signature concept have been suggested. However, they suffer from different drawbacks, especially the security relies on the assumption, that all t verifiers must be *honest* during verification. In this paper, these weaknesses are pointed out and a new scheme is suggested. Furthermore, new protocols for the two other concepts are presented. Our solutions demonstrate the close relation between signature schemes with threshold verification and threshold cryptosystems.

1 Introduction

In the conventional digital signature concept one signer is needed to generate a digital signature which can be verified by any verifier. In multiparty concepts for digital signatures, it is required that many signers sign a message together or only many verifiers can verify a given signature simultaneously. In this paper we will deal with the latter case. More precisely, we focus on signature schemes with multiparty-verification, where only a threshold of at least t out of n verifiers can verify a given signature. Depending on the fact, if these verifiers are predetermined or not by the signer in the group G of n potential verifiers, we obtain three classes of signature schemes with (t, n) shared verification.

1. *Signature with non-anonymous (t, n) shared verification*:
   During signature generation t verifiers of G are determined by the signer, such that only they are able to verify a given signature. Each possible group of t verifiers has a common key, that is derivated from the common group key.

---

[1] The author's work was granted by a postdoctoral fellowship of the NATO Scientific Committee disseminated by the DAAD. His current address is r³ security engineering, Zürichstrasse 151, CH-8607 Aathal, petersen@r3.ch.

2. *Signature with anonymous (t, n) shared verification:*
   The signer predetermines a group G of n persons, which share a common group key. During signature generation, the signer doesn't fix, which t verifiers in G are able to verify a given signature, thus they remain anonymous to the signer. Less than t verifiers are unable to verify the signature.
3. *Signature with convertible non-anonymous (d, t, n) shared verification:*
   At signature generation the signer determines $d \geq t$ verifiers of G, which are able to verify a signature. If he wants to convert this directed signature into an anonymous signature with (t, n) shared verification, he can do this by releasing some additional information. It is distinguishable, if the ability of a conversion is *verifiable* for the d verifiers in advance or if it is *voluntary* by the signer and only verifiable after conversion.

The case $d < t$ violates the property that for an anonymous (t, n) shared verification at least t verifiers are required and is therefore not considered. There exists a close relation between these signature concepts and threshold cryptosystems [Desm87, DeFr90]. We will show, how each concept can be realized using these schemes as a tool.

Applications for these classes of signature schemes are all situations, where the concurrency of all verifiers is important, such that no verifier can gain an advantage of the knowledge of the authentic message before the other group members know its validity (e.g. tenders, bits).

So far, there have been only proposals for the second concept by de Soete, Quisquater and Vedder [SoQV89], Harn [Harn93] as well as Lim and Lee [LiLe96]. The first scheme is based on geometry and not resistant against dishonest verifiers. Harn's scheme was broken by Lee and Chang [LeCh95]. Then, two modifications have been proposed by Harn [Harn95], one of which was also insecure and has been improved [HMP95a] and the other one needed an unreasonable amount of storage. However, a problem of the improved scheme and the scheme by Lim and Lee is that the verifiers have to be assumed to be honest, which might not be the case in practice.

In the following, we present solutions for the three classes, where malicious verifiers can be detected and identified. For the second class, we also review one scheme and point out its vulnerability to dishonest verifiers.

2   General aspects

A group G consists of n users $U_1, \ldots, U_n$ with identities $ID_1, \ldots, ID_n$. To analyze the security of signature schemes with (t, n) shared verification, we have to consider the following aspects:

1. Security of the signature against forgery.
2. Necessity of at least t verifiers from G for (first) verification.
3. Disruption of the verification by malicious verifiers.

After a first verification, it is allowed, that the signature is checked by anyone else, but this person should be aware, that the signature was initially generated for the group G. Additionally, in the case of *non-anonymous* signature verification the aspect that only the intended group of d users is able to verify the signature (for the first time) has to be considered. Furthermore, it has to be guaranteed that the d verifiers can't modify the digital signature after verification such that it fits for the verification by another (innocent) group of d' verifiers, who doesn't know that the signed message was intended by the signer for someone else. A stronger property would be, that the directed signature is not transferable to anyone else also after a first verification, as in the concept of undeniable signatures [ChAn89]. This property is not considered here.

In the case of *convertible schemes* also the aspect that the conversion is *consistent* has to be considered, namely that valid (invalid) signatures can only be converted into valid (invalid) signatures and that they can't change their status of validity by conversion.

As a tool, we use the non-interactive proof of equality of two discrete logarithms [ChPe92]:

Definition 1 (Proof$_{LOGEQ}$) *A (message-dependent) proof of equality of the discrete logarithm of y_1 to the base α_1 and y_2 to the base α_2 is a pair $(r, s) := Proof_{LOGEQ}(m, \alpha_1, y_1, \alpha_2, y_2)$ satisfying the following condition:*

$$r := H(m||\alpha_1||y_1||\alpha_2||y_2||\alpha_1^s \cdot y_1^r \ (mod \ p)||\alpha_2^s \cdot y_2^r \ (mod \ p)).$$

The message m can be the empty string ϵ.

This proof can be obtained, if and only if the prover knows the discrete logarithms $\log_{\alpha_1}(y_1)$ and $\log_{\alpha_2}(y_2)$ and they are both equal. To construct the proof, the prover chooses a random value $k \in_R Z_q$, calculates $r := H(m||\alpha_1||y_1||\alpha_2||y_2||\alpha_1^k \ (mod \ p)||\alpha_2^k \ (mod \ p))$ and $s := k - r \cdot x \ (mod \ q)$. It is assumed, that this proof doesn't leak any useful information about x. The proof can be extended to show the equality of $n \geq 2$ discrete logarithms in an obvious manner. This will be abbreviated by $Proof_{LOGEQ}^n$ in the following.

The security of some computation relies on the *Extended Diffie-Hellman problem* which we assume, is not solvable in probabilistic polynomial time as it is equivalent to the basic Diffie-Hellman problem.

Definition 2 (Extended Diffie-Hellman Problem) *Given a prime modulus p and a primitive root $\alpha \in Z_p^*$. The Extended Diffie-Hellman problem is to compute $K := \alpha^{b \cdot \sum_{i=1}^{n} a_i} \ (mod \ p)$ from the knowledge of $B := \alpha^b \ (mod \ p)$ and the n values $A_i := \alpha^{a_i} \ (mod \ p)$, $i \in [1 : n]$.*

Obviously, the special case $n = 1$ is exactly the Diffie-Hellman problem. Thus the Diffie-Hellman problem is polynomial time reducible to the Extended Diffie-Hellman problem. On the other hand, if the Diffie-Hellman problem is solvable, it is possible to compute all n values $\alpha^{b \cdot a_i}$ and obtain the value K as a product of these n values.

In the following, we use small characters for parameters that are generated by single users while capital characters are used for shared values, that can be generated only by a group of t users. The function h is choosen as a suitable hash function which is publicly known to all users.

3 Non-anonymous threshold verification

We propose to use a directed signature [LiLe93], where each user has his own private and public key pair. Thus the signature σ can be encrypted, such that only the selected t users can decrypt it and verify the signature hereafter. It is sufficient to use an encryption scheme with verifiable correct multi-decryption in combination with an ordinary signature scheme from the family of ElGamal signature schemes [ElGa85, HoMP94]. We demonstrate the concept using the ElGamal encryption and the Schnorr signature scheme [ElGa85, Schn91].

3.1 The new scheme

Initialisation
The certification authority generates two primes p, q with $q|(p-1)$ and a generator α of order q. She also publishes a suitable hash function h.

Key generation
Each user U_i chooses a random secret key $x_i \in_R \mathbf{Z}_q$ and generates his public key $y_i := \alpha^{x_i} \pmod{p}$ which is certified by the certification authority.

Signature generation
We assume, that the verifiers $U_1, \ldots, U_t$ should verify the multi-signature. To sign a message m, Alice chooses a random number $k \in_R \mathbf{Z}_q$, computes $c := \alpha^k \pmod{p}$, $R := \prod_{i=1}^{t} y_i^k \pmod{p}$ and $s := k - x_A \cdot h(R, m) \pmod{q}$. The tuple (c, s) is the directed multi-signature for the message m.

Signature verification
To check the validity of the signature (c, s) of message m, verifier U_i $(1 \le i \le t)$ calculates $r_i := c^{x_i} \pmod{p}$ and $z_i := y_A^{x_i} \pmod{p}$. Furthermore he computes $V_i := Proof^3_{LOGEQ}(m, \alpha, y_i, c, r_i, y_A, z_i)$, which is only possible, if the values r_i and z_i are computed as described above. The values r_i, z_i, V_i are sent via a broadcast channel to the other $t - 1$ verifiers, which check all proofs $V_i, i \ne j$ and compute

$$R := \prod_{i=1}^{t} r_i \pmod{p} \quad \text{and} \quad Y := \prod_{i=1}^{t} z_i \pmod{p}.$$

Now, the signature is verified using the equation:

$$R \equiv \left(\prod_{i=1}^{t} y_i\right)^s \cdot Y^{h(R,m)} \pmod{p}. \tag{1}$$

If the same group of verifiers check many signatures of the same signer, they can store the value Y and don't have to recompute it every time. This results in more efficient proofs $V_i := Proof_{LOGEQ}(m, \alpha, y_i, c, r_i)$ for the correct computation of r_i as the parameters z_i don't have to be considered anymore. Thus the signature verification gets more efficient.

3.2 Security considerations

1. *Security against forgery:* The security of the signature scheme relies on the security of the Schnorr signature scheme with public parameters p, q, generator $\prod_{i=1}^{t} y_i$ and signer's key pair x_A, $Y := \prod_{i=1}^{t} y_i^{x_A} \pmod{p}$.
2. *Necessity of all t verifiers for verification:* Without the knowledge of the values R and Y, the verification equation (1) can't be verified. As a result any unauthorized group G of malicious verifiers can't generate the values R and Y themselves. This is equivalent to the Extended Diffie-Hellman problem with parameters $(c, \prod_{i=1}^{t} y_i)$ or $(y_A, \prod_{i=1}^{t} y_i)$. Thus the knowledge of all t secret values x_i is required. Even one of the authorized verifiers who already knows the value Y from a previous computation can't check the validity of the signature without the help of the other $t - 1$ verifiers, as he can't compute the (random) value R out of c himself. Thus, only the intended verifiers together can verify a signature
3. *Disruption by a malicious verifier:* As the correct computation of all values r_i and z_i is verified by the proofs V_i, no verifier is able to cheat undetectable.
4. *Separation of signature and encrypted values:* The signature parameter R is Diffie-Hellman encrypted with multi-decryption by the t designated verifiers. To re-encrypt the value R with the public key of another group of v verifiers $U_{\pi(1)}, \ldots, U_{\pi(v)}$, the designated verifiers need to compute the discrete logarithm of R to the base $\prod_{i=1}^{v} y_{\pi(i)}$ in order to generate a valid ciphertext $\tilde{c}$. This would also break the signature scheme totally but is assumed not to be possible in polynomial time. Thus the encryption can't be separated from the signature.

4 Anonymous (t,n) shared verification

We first describe the scheme from [LiLe96] and mention Harn's scheme [Harn93, HMP95a]. The two schemes differ in the fact, that the first uses *dynamic* shares of a secret shared among the members of G which are distributed together with the signature, while the second uses *static* (pre-distributed) shares and is thus more efficient, as it decreases the size of each signature. However. both schemes are insecure under the assumption, that not all verifiers are honest. Therefore we present a new solution.

To show the correctness of the schemes and to simplify the notations, we need the following theorem [Sham79]:

Theorem 1 (Lagrange) *Let q be a prime, $x, a_1, \ldots, a_{t-1} \in_R \mathbf{Z}_{q-1}$ and the function f be equal to $f(u) := x + a_1 \cdot u + a_2 \cdot u^2 + \ldots + a_{t-1} \cdot u^{t-1} \pmod{q}$. Further, the values B_j are given as $B_j := \prod\limits_{z=1, z \neq j}^{t} \frac{-ID_j}{ID_j - ID_z} \pmod{q}$. Then the following equation holds:*

$$\sum_{j=1}^{t} f(ID_j) \cdot B_j \equiv f(0) \equiv x \pmod{q}$$

The variable B_j is used in the following to clarify the notations.

4.1 The Lim and Lee scheme

We use a variant of the Schnorr signature scheme [Schn91] for designated verifiers [HMP95c, LiLe96]. The initialization and key generation is done as in the last section.

Signature generation
To sign a message m, the user Alice chooses $k, X \in_R \mathbf{Z}_q$ and computes $Y := \alpha^X \pmod{p}$ and $R := Y^k \equiv \alpha^{X \cdot k} \pmod{p}$. The value X serves as a one-time encryption key for the signature parameter R. Then she computes $c := h(Y, R, m)$ and $s := k - x_A \cdot c \pmod{q}$. She randomly chooses the values $a_1, \ldots, a_{t-1} \in_R \mathbf{Z}_{q-1}$ and defines the function f as $f(u) := X + a_1 \cdot u + \ldots + a_{t-1} \cdot u^{t-1} \pmod{q}$. Alice calculates the n shadows $S_i := f(ID_i) \cdot y_i^X \pmod{p}$, where ID_i is the identity of user U_i and y_i his public key. The values S_i are the ElGamal encrypted shares $f(ID_i)$ under the secret keys x_i. The values S_i are broadcasted together with the signature (Y, c, s, m).

Signature verification
Each verifier U_i recovers $f(ID_i) = S_i \cdot Y^{-x_i} \pmod{p}$. Then he computes

$$r_i := (\alpha^s \cdot y_A^c)^{B_i \cdot f(ID_i)} \pmod{p}$$

and sends r_i to the group center Z. Z computes $R := \prod_{i=1}^{t} r_i \pmod{p}$ and verifies the signature by checking, that $c = h(Y, R, m)$.

Cryptanalysis

The t verifiers must be *honest*. Otherwise there exist three attacks:

1. A malicious verifier U_i could avoid the successful verification of the signature by giving a false value r_i to the group center Z.
2. More seriously, a malicious verifier can *universally forge* a signature on the message m and convince the group center Z, that it is a valid signature of the signer Alice. This is done, by choosing Y and all r_i at random and computing $c := h(Y, \prod_{i=1}^{t} r_i \pmod{p}, m)$. Thus Z will be convinced by the validity of the signature.

3. If the group center and a malicious verifier (e.g. verifier U_1) collude, they are able to convince the other verifiers that a signature forged by themselves is true: They just pick random X, R and compute $Y := \alpha^X \pmod{p}$ and $c := h(Y, R, m)$. They further compute S_i as described for the signature generation (it should be noted that the secret key is not needed for this computation). During the verification process, each honest verifier U_i computes $f(ID_i)$ and r_i as described and sends r_i to the group center Z. The malicious verifier U_1 waits until the group center knows all values r_i. Then he (or the group center) computes $r_1 := R/(\prod_{i=2}^{t} r_i)$. As a result, the verification equation is satisfied.

4.2 The modified Harn scheme

It can be shown that the modified Harn's scheme [Harn93, HMP95a], which is secure against universal forgery, suffers also from the first weakness pointed out above. It seems possible to repair this scheme by adding proofs of correct computation. The description is skipped because of space limitations.

4.3 The new scheme

A quite simple approach is to sign a message using a signature scheme and distribute a suitable parameter using a secret sharing scheme following Franklin and Reiter [FrRe95]. However, the disadvantage of this approach is that the length of the signature depends on the number of group members and is not fixed. This drawback can be avoided by using an encryption scheme with provably correct threshold decryption. Then the threshold signature scheme can be realized by encrypting one of the signature parameters. Such a scheme can be realized by adding the property of verifiable decryption to the threshold decryption scheme given in [DeFr90].

The verifiable threshold decryption scheme

We describe a verifiable threshold decryption scheme based on the ElGamal encryption scheme [ElGa85, DeFr90] and add the feature that decryption is proven to be correct.

1. *Initialization:* The certification authority chooses large primes p, q with $q|(p-1)$ and a generator $\alpha \in_R \mathbf{Z}_p^*$ of order q.
2. *Key generation:* The certification authority chooses a secret key $x_G \in_R \mathbf{Z}_q^*$, the values $a_1, \ldots, a_{t-1} \in_R \mathbf{Z}_q$ and defines the function f as $f(u) := x_G + a_1 \cdot u + \ldots + a_{t-1} \cdot u^{t-1} \pmod{q}$. It transmits the value $f(ID_i)$ to the user U_i secretly. Furthermore, it publishes the certified tuples $(ID_i, S_i := \alpha^{f(ID_i)} \pmod{p})$ as reference values and the public group key $y_G := \alpha^{x_G} \pmod{p}$. The values $x, a_1, \ldots, a_{t-1}$ are kept secret.

3. *Encryption:* To encrypt a message $m \in \mathbf{Z}_{p-1}$, sender Alice picks a random number $k \in_R \mathbf{Z}_q^*$, computes $c_1 := \alpha^{-k} \pmod{p}$ and $c_2 := m \cdot y_G^k \pmod{p}$. Alice publishes the tuple (c_1, c_2). The encryption is abbreviated by $(c_1, c_2) := E(y_G, m)$.

4. *Decryption:* Every receiver U_i (assume here, that these are the receivers $U_1, \ldots, U_t$) computes the value $K_i := c_1^{f(ID_i)B_i} \pmod{p}$ and the proof $V_i := \mathrm{Proof}_{LOGEQ}(m, \alpha, S_i, c_1^{B_i}, K_i)$. The values K_i and V_i are send via a broadcast channel to the other $t-1$ receivers, such that all receivers U_j can verify the proofs $V_i, i \neq j$ and compute $K := \prod_{i=1}^{t} K_i \pmod{p}$ and $m := K \cdot c_2 \pmod{p}$. The decryption is abbreviated by $m := D(c_1, c_2, f(ID_1), \ldots, f(ID_t))$.

If the users don't trust the certification authority that she honestly distributes the shares $f(ID_i)$, it is possible to replace the secret sharing scheme for the secret x_G in the initialization by the verifiable secret sharing scheme (VSS) suggested in [Ped91a]. In this case, the authority broadcasts the values $\alpha^{a_i} \pmod{p}$, $i \in [1 : t-1]$ to all users, such that everyone can verify his share $f(ID_i)$ by

$$\alpha^{f(ID_i)} \equiv y_G \cdot \prod_{j=1}^{t-1} (\alpha^{a_j})^{ID_i^j} \pmod{p}.$$

Furthermore, it is also possible that the users generate their shared secret themselves without the help of a trusted party, as described in [Ped91b]. This eliminates the trust in a certification authority for the cost of a higher computational effort.

The signature scheme

As an example we use the Schnorr signature scheme [Schn91]. The initialization and key generation is the same as in the verifiable threshold decryption scheme. Additionally, a hash function h is chosen.

Signature generation

To sign m, Alice chooses a random number k, generates $r := \alpha^k \pmod{p}$ and computes $s := -x_A \cdot h(r, m, desc(G)) + k \pmod{q}$, where $desc(G)$ is a suitable description of the group G. Then she encrypts the signature parameter s using the public group key y_G as $(c_1, c_2) := E(y_G, s)$. The tuple (r, c_1, c_2, m) is published as signed message.

Signature verification

To check the validity of the signature (r, c_1, c_2), t verifiers out of the group G first decrypt $s := D(c_1, c_2, f(ID_1), \ldots, f(ID_t))$ and then verify the Schnorr signature (r, s) by the equation $r \equiv \alpha^s \cdot y_A^{h(r, m, desc(G))} \pmod{p}$.

Security considerations

1. *Security against forgery:* The security of the scheme relies on the security of Schnorr's signature scheme.

2. *Necessity of all t verifiers for verification:* If less than t verifiers collaborate, they can't compute K, as the secret values $f(ID_i)$ are needed for this computation. As a result, s can't be decrypted and the verification equation can't be checked. The knowledge of an old signature parameter s doesn't give any advantage for the verification, as each individual signature consists of a new signature parameter s, which is not related to formerly chosen ones. If a decrypted signature (r, s) is shown to someone else, he will be aware, that it was originally signed for the group G, as a group description is embedded in the hash function.

3. *Disruption by a malicious verifier:* During the decryption of s it's impossible for a malicious verifier U_j to deviate from the protocol, as otherwise he is unable to give a correct proof V_j. V_j demonstrates, that there exists a value z, such that $\alpha^z \equiv S_i \pmod{p}$ and $c_1^{B_1 \cdot z} \equiv K_i \pmod{p}$. As the discrete logarithm of the certified value S_i is known to be $f(ID_i)$ to the base α, it implies, that $K_i \equiv c_1^{B_1 \cdot z} \pmod{p}$ holds.

5 Convertible threshold verification

In a convertible signature scheme with non-anonymous threshold verification, d verifiers are predetermined by the signer and the signature is generated accordingly. Later, the signer is able to convert this signature by verifiable releasing some previously secret information, such that any $t \leq d$ out of n members of group G can verify it. The signer has two possibilities to do this conversion:

1. *Selective conversion:* Only a single non-anonymous threshold signature is converted into an anonymous threshold signature.

2. *Total conversion:* All non-anonymous threshold signatures for a specific group $\bar{G}$ of d verifiers are converted into an anonymous (t, n) shared verification signatures.

To obtain a *convertible* signature scheme, Alice signs the message m using the proposed signature scheme from section 4.3, where the signature parameter s is encrypted as $E(y_G, s)$. Additionally, she encrypts the signature parameter r using an encryption scheme with verifiable (d, d)-threshold decryption, e.g. by using the scheme proposed in [HMP95b]. Thus only the d designated verifiers are able to decrypt the signature parameter r and verify the signature.

Alice selectively converts the directed signature by releasing the encrypted parameter r and proving, that r was really encrypted. For total conversion, she generates an additional key pair. She encrypts r using the public key of that additional key pair as well and proves this. Then, the related secret key is released for total conversion. To avoid separation of the signature and the ciphertext, one parameter of the encryption scheme is embedded into the hash of the original message.

5.1 Convertible signature scheme with threshold verification

We describe the scheme using the signature scheme with anonymous threshold verification. The initialization and the key generation is the same as in section 4.3. Additionally, each signer creates a key pair (X, Y) for each possible group $\bar{G}, |\bar{G}| = d$, of designated verifiers for total conversion. He proves to the verifiers, that he knows X to obtain a verifiable convertible scheme.

Signature generation
We assume w.l.o.g. that the designated verifiers are $U_1, \ldots, U_d$. Alice picks a random k and computes $c_3 := \alpha^{-k} \pmod{p}$. Then she generates a signature (r, c_1, c_2) on m and c_3 for the group $\bar{G}$ using the signature scheme from section 4.3. After this, she computes $y_{\bar{G}} := \prod_{i=1}^{d} y_i \pmod{p}$, $c_4 := y_{\bar{G}}^k \cdot r \pmod{p}$, $c_5 := Y^k \cdot r \pmod{p}$ accordingly. Additionally, she calculates $C := \text{Proof}_{LOGEQ}(\epsilon, \alpha^{-1}, c_3, y_{\bar{G}}/Y \pmod{p}, c_4/c_5 \pmod{p})$. $(c_1, c_2, c_3, c_4, c_5, C)$ is the convertible signature with threshold verification on message m.

Signature verification
To check a signature $(c_1, c_2, c_3, c_4, c_5, C)$ on a message m, verifier U_j $(1 \leq j \leq d)$ checks that C is valid, calculates $z_j := c_3^{x_j} \pmod{p}$ and $V_j := \text{Proof}_{LOGEQ}(m, \alpha, y_j, c_3, z_j)$. The values z_j, V_j are sent to the other $d - 1$ verifiers via a broadcast channel, who check the correctness of V_j. Then $r := c_4 \cdot \prod_{j=1}^{d} z_j \pmod{p}$ is computed by every verifier. Finally the signature (r, c_1, c_2) on message m is checked as described in section 4.3.

Conversion of a signature
After conversion of a signature, the proof C is still checked. This guarantees the anonymous verifiers, that the signature was converted consistently.

- To *selective convert* a valid signature, Alice releases the signature parameter r together with $\text{Proof}_{LOGEQ}(\epsilon, \alpha^{-1}, c_3, y_{\bar{G}}, c_4 \cdot r^{-1} \pmod{p})$. Thus any user can verify this proof and use r as signature parameter in the signature with anonymous threshold verification.
- To *totally convert* all signatures, Alice releases the secret key X, such that every user can compute all values $r := c_3^X \cdot c_5 \pmod{p}$. Then each signature can be verified by any t users in the group G.

5.2 Security considerations

The security of the convertible signature depends on the security of the underlying threshold signature scheme, which has been discussed in section 4.3. Its verification is assumed to be possible only after the successful decryption of the signature parameter r by the d designated verifiers.

1. *Unforgeability:* To forge a converted threshold signature, an attacker must be able to generate $(c_1, c_2, c_3, c_4, c_5, C)$ on a message m such that

$r \equiv \alpha^s \cdot y_A^{h(r,m,desc(G))} \pmod{p}$, $s = D(c_1, c_2, f(ID_1), \ldots, f(ID_t))$, $r \equiv c_4/c_3^{\sum_{i=1}^{d} x_i} \pmod{p}$ and proof C holds. Obviously, if such a signature can be generated by an attacker who colludes with d verifiers, then Schnorr's signature scheme could be broken as well.

2. *Necessity of all verifiers in $\bar{G}$ for verification:* It's a consequence of the used sharing technique, that d (or t in the converted case) signers are needed to decrypt the encrypted parameter r (or s), provided the Diffie-Hellman assumption holds.

3. *Disruption by a malicious verifier:* A verifier U_j cannot disrupt the protocol by computing a wrong value z_j, as otherwise he could not give a correct proof V_j.

4. *Separation of signature and encryption:* As the encryption parameter c_3 is signed, signature and encryption cannot be separated. To re-encrypt r for another group H with public key y_H using the same $c_3 \equiv \alpha^{-k} \pmod{p}$, one has to solve the Diffie-Hellman problem for c_3^{-1} and y_H to successfully generate $c_4 := y_H^k \cdot r \pmod{p}$.

5. *Consistent conversion:* The selective converted signature is consistent as the signer proves that he reveals the correct parameter r by proof $Proof_{LOGEQ}$ and checks C. Regarding total conversion, the proof C holds for all valid signatures. It demonstrates, that the same parameter r is encrypted in c_5 under the public key $y_{\bar{G}}$ and in c_4 under the public key Y. Therefore the property is satisfied.

6   Conclusion

In this paper we presented schemes for different signature concepts with (t, n) shared verification. Furthermore we have pointed out some limitations of two known solutions to one of the concepts. The proposed schemes can easily be combined with proposals for signature schemes with (t, n) shared signature generation, as presented by Park and Kurosawa [PaKu96], Gennaro et al. [GJKR96].

References

[ChAn89] D.Chaum, H. van Antwerpen, "'Undeniable Signatures'", LNCS 435, Advances in Cryptology: Proc. Crypto '89, Springer, (1990), pp. 212 – 216.

[ChPe92] D.Chaum, T.Pedersen, "Wallet databases with observers", LNCS 740, Advances in Cryptology: Proc. Crypto'92, Springer, (1993), pp. 89 – 105.

[Desm87] Y.Desmedt, "Society and group oriented cryptography : a new concept", LNCS 293, Advances in Cryptology: Proc. Crypto'87, Springer, (1988), pp. 120 – 127.

[DeFr90] Y. Desmedt, Y. Frankel, "Threshold cryptosystems", LNCS 537, Advances in Cryptology: Proc. Crypto '90, Springer, (1991), pp. 307–315.

[ElGa85] T.ElGamal, "A public key cryptosystem and a signature scheme based on discrete logarithms", IEEE Transactions on Information Theory, Vol. IT-30, No. 4, July, (1985), pp. 469 – 472.

[FrRe95] M.K.Franklin, M.K.Reiter, "Verifiable signature sharing", LNCS 921, Advances in Cryptology: Proc. Eurocrypt'95, Springer, (1996), pp. 50 – 63.

[GJKR96] R.Gennaro, S.Jarecki, H.Krawczyk, T.Rabin, "Robust Threshold DSS Signatures", Advances in Cryptology: Proc. Eurocrypt'96, LNCS 1070, Springer, (1996), pp. 354–371.

[Harn93] L.Harn, "Digital signature with (t, n) shared verification based on discrete logarithms", Electronics Letters, Vol.29, No. 24, (1993), pp. 2094–2095.

[Harn95] L.Harn, Reply to [LeCh95], Electronics Letters, Vol.31, No. 3, (1995), pp. 177.

[HoMP94] P.Horster, M.Michels, H.Petersen, "'Meta-ElGamal signature schemes"', Proc. 2. ACM conference on Computer and Communications security, ACM Press, November, (1994), pp. 96 – 107.

[HMP95a] P.Horster, M.Michels, H.Petersen, "Comment: Digital signature with (t, n) shared verification", Electronics Letters, July, (1995), pp. 1137.

[HMP95b] P.Horster, M.Michels, H.Petersen, "Blind multisignature schemes and their relevance to electronic voting", Proc. 11th Annual Computer Security Applications Conference, IEEE Press, December, (1995), pp. 149–156.

[HMP95c] P.Horster, M.Michels, H.Petersen, "'Das Meta-ElGamal Signaturverfahren und seine Anwendungen"', Proc. VIS' 95, Rostock, DuD Fachbeiträge, Vieweg Verlag, (1995), pp. 207 – 228.

[LeCh95] W.-B.Lee, C.-C.Chang, "Comment: Digital signature with (t, n) shared verification based on discrete logarithms", Electronics Letters, Vol.31, No.3, (1995), pp. 176–177.

[LiLe93] C.H.Lim, P.J.Lee, "Algorithmic measures for preventing middleperson attack in identification schemes", Electronics Letters, Vol. 29, No. 14, (1993), pp. 1281 – 1282.

[LiLe96] C.H.Lim, P.J.Lee, "Directed Signatures and Applications to Threshold Cryptosystems", LNCS 1189, Workshop on Security Protocols, Springer, (1997), pp. 131 – 138.

[PaKu96] C.Park, K.Kurosawa, "New ElGamal Type Threshold Digital Signature Scheme", IEICE Trans. Fundamentals. Vol. E79-A, No. 1, January, (1996), pp. 86–93.

[Ped91a] T.Pedersen, "Distributed Provers with Applications to Undeniable Signatures", LNCS 547, Advances in Cryptology: Proc. Eurocrypt'91, Springer, (1992), pp. 221 – 238.

[Ped91b] T.Pedersen, "A threshold Cryptosystem without a Trusted Party", LNCS 547, Advances in Cryptology: Proc. Eurocrypt'91, Springer, (1992), pp. 522 – 526.

[PoSt96] D.Pointcheval, J.Stern, "Security Proofs for Signatures", LNCS 1070, Advances in Cryptology: Proc. Eurocrypt'96, Springer, (1996), pp. 387 – 398.

[Schn91] C.P.Schnorr, "Efficient signature generation by smart cards", Journal of Cryptology, Vol. 4, (1991), pp. 161–174.

[Sham79] A. Shamir, "How to share a secret", Communications of the ACM, Vol. 22, No. 11, November, (1981), pp. 612–613.

[SoQV89] M.de Soete, J.J.Quisquater, K.Vedder, "A signature with shared verification scheme", LNCS 435, Advances in Cryptology: Proc. Crypto'89, Springer, (1990), pp. 253 – 262.

Open Key Exchange:
How to Defeat Dictionary Attacks Without Encrypting Public Keys

Stefan Lucks

Institut für Numerische und Angewandte Mathematik
Georg–August–Universität Göttingen
Lotzestr. 16–18, D–37083 Göttingen, Germany
(email: lucks@math.uni-goettingen.de)

Abstract. Classical cryptographic protocols based on shared secret keys often are vulnerable to key-guessing attacks. For security, the keys must be strong, difficult to memorize for humans. Bellovin and Merritt [4] proposed "encrypted key exchange" (EKE) protocols, to frustrate key-guessing attacks. EKE requires the use of asymmetric cryptosystems and is based on encrypting the public key, using a symmetric cipher.

In this paper, a novel way of key exchange is presented, where public keys are sent openly, not encrypted. In contrast to EKE protocols, the same public-key/secret-key pair can be used for arbitrary many protocol executions. The RSA-based protocol variant is found to be quite efficient and practical.

Compared to previous work on such protocols, a more solid formal treatment is given, influenced by the work of Bellare and Rogaway [3] on key exchange protocols for *strong* common secrets.

1 Introduction

Most humans, including the current author, are quite bad at memorizing large randomly generated keys. In the real world, many people choose weak passwords or use a short PIN ("personal identification number"). Hence, there is a real need for cryptographic protocols to give a reasonable degree of security even if rather weak keys are used.

In 1992, Bellovin and Merritt [4] (see also section 22.5 of [12]) addressed this problem. Their solution was an "encrypted key-exchange" (EKE) protocol. The core idea behind EKE is to use the secret (symmetric) key π to encrypt the public key E of a randomly chosen asymmetric key pair (E, D), then to use E to encrypt the randomly chosen (symmetric) session key k and finally to decrypt $E(k)$ using D. The security of EKE depends on the fact that two parties using different values for π are likely to use different public keys E and E', and that knowing an asymmetric key pair (E', D') does not help to decrypt $E(k)$. It is important for EKE protocols, that every public key is used only once.

This paper's approach is somewhat different from EKE. Though we also make use of public key cryptography, the public key is not encrypted but sent openly in the clear. Also, the public key may be reused arbitrarily often.

2 Our Model of Communication

In this section[1], we formalize how Alice and Bob exchange keys while the adversary Eve controls the communication network. Our model is inspired by the work of Bellare and Rogaway [3]. At her will, Eve can deliver messages between Alice and Bob. She is able

- to read all messages between both parties,
- to modify such messages, to delay them or to send them multiple times,
- and to send messages generated by herself to Alice or Bob.

Also, Eve can create several instances of Alice and Bob and communicate with these instances in parallel sessions. Both Alice and Bob are represented by an infinite collection of oracles. These oracles can be called by Eve and then change their internal state and compute an output, but they never directly communicate with each other. We denote the oracles by Π_A^i resp. Π_B^j. A and B indicate "Alice" resp. "Bob" while i and $j \in \mathbb{N}$ indicate the instance ($\mathbb{N} = \{1, 2, \ldots\}$).

The functions Π_A and Π_B define the behavior of Alice and Bob. For $I \in \{A, B\}$, the functions Π_I takes the following inputs:

- 1^K (the security parameter $K \in \mathbb{N}$ in unary notation),
- $\pi \in \{0, 1\}^*$ (the secret information),
- $c \in \{0, 1, -\}^*$ (the communication so far; "$-$" is used as a separator),
- and $r \in \{0, 1\}^\infty$ (an infinite amount of fair private coin flips),

to compute the outputs pair (x, y) with

- $x \in \{0, 1\}^*$ (the next message to send—the "reply")
- and $y \in \{\{0, 1\}^*, \text{reject}, \text{none}\}$.

The oracle can accept a communication, then $y \in \{0, 1\}^*$ is the session key. Also, the oracle can reject a communication or refuse to make a decision at that point of time. If it accepts or rejects, the decision is final.

Running a protocol in the presence of Eve, given a security parameter K, means applying the following experiment:

1. Choose a secret π.
2. For $I \in \{A, B\}$ and $j \in \mathbb{N}$, initialize the $c_{I,j} \in \{0, 1\}^*$ by the empty string.
3. For $I \in \{A, B\}$ and $j \in \mathbb{N}$, choose a random string $r_{I,j} \in \{0, 1\}^\infty$.
4. Run the adversary Eve, answering oracle calls as follows. When Eve asks a query $(I, j, z) \in \{A, B\} \times \mathbb{N} \times \{0, 1\}^*$, oracle Π_I^j

    - replaces $c_{I,j}$ by the concatenation of $c_{I,j}$, "$-$", and z,
    - computes $(x, y) := \Pi_I \left(1^K, \pi, c_{I,j}, r_{I,j}\right)$,
    - replies x,
    - and tells Eve, whether it accepted, rejected, or did neither.

---

[1] Readers with little interest in the formal treatment of cryptographic protocols may skip ahead to the next section (and possibly go back to this section later).

Once the oracle Π_I^j has accepted, Eve can ask Π_I^j for its session key.

5. Finally, Eve chooses an oracle Π_J^i and attempts to guess its session key.

If Eve chooses a pair of oracles Π_A^i and Π_B^j and then faithfully conveys each flow from one oracle to the other, then we call the oracles Π_A^i and Π_B^j **related**. If two oracles are related, Eve did behave like a **benign adversary**[2].

The adversary Eve is **successful**, if it guesses the the session key of an accepting oracle Π_I^j without having asked either Π_I^j or a related oracle for its session key. Note that a benign adversary still can either ask accepting oracles for their session keys or attempt to guess their session keys.

In addition to the above, Alice and Bob, or their instances, reject after waiting a reasonable amount of time for an answer from their partner (time-out).

Now, at last, we can define what we are talking about in this paper:

Definition. Let π be randomly chosen from the set S_π of size S, according to the uniform probability distribution. We say that the protocol defined by the pair of functions Π_A and Π_B is a **secure key exchange protocols for weak common secrets**[3],

D1: if two oracles are related, then with overwhelming probability both accept and compute identical session keys,

D2: if the probability of success is negligible for every efficient benign adversary,

D3: and if, after Eve has been rejected R times, her probability of success without being rejected again does not significantly exceed $1/(S - R)$.

Condition D1 says, if each party's messages are faithfully relayed to one another, then the parties succeed in finding a common session key—at least with overwhelming probability. Due to condition D2, passive attacks are hopeless. Condition D3 appears somewhat complicated, but it simply means that active attacks cannot significantly improve on the trivial attack. The trivial attack consists of Eve choosing any value $\psi \in S_\pi$ for π and then masquerading as either Alice or Bob. If the other party accepts, Eve concludes that her guess was correct. If the other party rejects, Eve knows $\psi \neq \pi$ and the next time chooses $\psi' \neq \psi$ for π. At first, Eve's probability of success is $1/S$, and after $R < S$ failed attacks it has grown to $1/(S - R)$.

We could generalize the above definition such that π is randomly chosen from the set S_π, according to *any probability distribution*, as long as for every element ψ of S_π the probability $\text{prob}(\pi = \psi)$ is bounded by $\text{prob}(\pi = \psi) \leq 1/S$. For the sake of simplicity, we restrict ourselves to the uniform probability distribution.

---

[2] This notion is due to Bellare and Rogaway [3]. If our key exchange protocol is any good, two related oracles both accept and their session keys are equal—at least with overwhelming probability.

[3] In the sequel, we sometimes simply use "password" instead of "weak common secret".

3 The generic open key exchange (OKE) protocol

In this section, we describe the generic open key exchange protocol, define its security requirements and prove it to be secure.

Our notation and requirements can be described like this:

Let K be our security parameter, large enough that brute-forcing 2^K values is infeasible, the number S of choices for the secret small enough that brute-forcing S^2 values is still feasible (hence $S \ll 2^K$), and $l \geq 2K$. For the protocol, we need hash functions $h_i : \{0,1\}^* \longrightarrow \{0,1\}^l$ $(i \in \{1,2,3\})$, an asymmetric key pair (E, D), such that E describes the encryption operation $E : S_E \longrightarrow S_E$, where S_E is a finite group "$\Diamond$" denotes the group operation and "$\neg$" its inverse, and a hash function $H : \{0,1\}^* \longrightarrow S_E$. The group operation needs to be compatible with the encryption function, i.e. for all $x, y \in S_E$: $E(x)\Diamond E(y) = E(x\Diamond y)$. Also, the encryption function must be invertible, i.e. $D(E(x)) = x$ must hold for $x \in S_E$.

The group S_E can depend on the choice of E, as for RSA. The pair (E, D) and hence the group S_E can be chosen once by Alice and remains fixed during the experiment of running the protocol in the presence of Eve. D is kept secret. We use "|" to indicate the concatenation of bit strings.

Our protocol works like this:

0. *Both in advance:* Alice and Bob agree on a common secret $\pi \in_R \{1, \ldots, S\}$. *Alice in advance:* Alice generates a public-key/secret-key pair (E, D).
1. Alice chooses $m \in_R \{0,1\}^l$ and sends E and m to Bob.[4]
2. Bob chooses $\mu \in_R \{0,1\}^l$ and $a \in_R S_E$, computes $p = H(E|m|\mu|\pi)$ and $q = E(a)\Diamond p$, and sends μ and q to Alice.
3. Alice computes p like Bob, $a' = D(q\neg p)$, and sends $r = h_1(a')$ to Bob. (Now $a' = a$.)
4. If $r \neq h_1(a)$, Bob rejects. Otherwise, he computes $k = h_2(a)$ and $t = h_3(a)$, sends t to Alice, and uses k as his session key.
5. Alice rejects if $t \neq h_3(a')$. Else, she uses $k' = h_2(a')$ as her session key.

We make the following two cryptographic assumptions:

A1: The hash functions H, h_1, h_2, and h_3 behave like independent random functions. In the notion of Bellare and Rogaway [2], the functions h_i are public "random oracles", i.e. Eve can send queries $x_1, x_2, \ldots$ to the random oracle h_i resp. H and receives answers $h_i(x_j)$, resp. $H(x_j)$, all independent random values. Only when a query $x_c = x_j$ is sent to the oracle h_i (or H) and Eve was given the oracle response $h_i(x_j)$ ($H(x_j)$) before, the result is predictable: $h_i(x_c) = h_i(x_j)$ (or $H(x_c) = H(x_j)$).

---

[4] Note that though the pair (E, D) can be chosen once, it does not make sense to assume Bob can remember E. If Bob already did know Alice's authentic public key E, he could simply choose the session key k_B and send $E(k_B)$ to Alice, instead of runing the protocol.

A2: Given a value x, randomly chosen from the set S_E, it is infeasible to find a value y with $y = D(x)$.

Note that A1 can actually be satisfied by using one single random function $h : \{0,1\}^* \longrightarrow \{0,1\}^l$, defining $h_1(x) = h(01|x)$, $h_2(x) = h(10|x)$, $h_3(x) = h(11|x)$, and H depending on the values $h(0000\ldots00|x)$, $h(0000\ldots01|x)$, $\ldots$, $h(00\ldots|x)$.

Theorem 3.1. *Under the assumptions A1 and A2, the generic OKE protocol is a secure key exchange protocol for weak common secrets.*

Throughout the proof, "essentially" means "with overwhelming probability".

Proof. One can easily verify that condition D1 holds.

The values m, μ, and a are independent random values from $\{0,1\}^l$ resp. S_e. Due to $D(E(a)) = a$, the attacker gets no information about p from $q = E(a)\diamond p$.

Given μ and q, Alice replies $r = h_1(a')$. If Eve did guess a', she could use r to check the correctness of this guess—and then compute π. But finding a' is equivalent to solving the equation $a' = D(q\neg p)$ and hence to decrypting a random ciphertext—even if p is known. Due to assumption A2, this is infeasible. By definition, the value $r = h_1(a')$ is just a random value for everyone not knowing a'. The same holds for $t = h_3(a)$. In other words, as long as Eve remains passive, all oracle replies look like random values from $\{0,1\}^l$ resp. from S_E for her—essentially. Thus, D2 is satisfied.

We still have to proof D3. Due to $l \geq 2K$, all instances Alice and Bob will create new values m, μ, and a, essentially.[5] This rules out replay attacks, where Eve waits for Alice to repeat herself to repeat Bobs responses, or for Bob to repeat himself, to repeat Alices responses.

Note that after starting an active attack, Eve must give a correct answer to avoid being rejected.

At first, we consider the case $R = 0$, i.e. no oracle did reject before when Eve attempts to guess a session key. Depending on the protocol step when Eve first becomes active, we argue:

1st step: If Eve actively chooses a pair (E, m) and sends it to Bob, she has to choose a response $r = h_1(a')$ in the third step with $a' = D(q\neg p)$. There are only S possible values p, thus there are essentially S different responses r. If Eve's choice is wrong, Bob rejects and Eve may be able to delete one of the S possibilities for p from her list. The probability for Eve's choice to be correct is $1/S$.

2nd step: If Eve, after having seen a pair (E, m) from an instance of Alice, actively chooses μ and q as a response, she receives the value $r = h_1(D(q\neg p))$ from Alice—and in principle knows enough to verify which value p is used

---

[5] We required K to be large enough that computing 2^K values is infeasible. But due to the birthday paradox, we'd need to call our oracles about $2^{l/2} \geq 2^K$ times to get one value m or μ which was used before.

by Alice. But due to assumption A1, the value r does Eve not help with computing a, other than as a verifyer whether the correct a has been found. Let Eve be able to find values a_1, p_1, a_2, p_2, and q with $p_1 \neq p_2$ two different candidates for p and

$$E(a_1)\Diamond p_1 = q = E(a_2)\Diamond p_2. \tag{1}$$

This means, she can find a plaintext $a_\Delta = a_1 \neg a_2$ with

$$E(a_\Delta) = p_2 \neg p_1,$$

and thus is able to decrypt the difference $p_2 \neg p_1$ of two random values. Since S is "small" and there are only S^2 such differences, this means that there is an efficient probabilistic algorithm for Eve to decrypt random values, in violation of assumption A2.

Thus, Eve can't know more than one candidate for a and hence not check more than one candidate for p. If her candidate for a is wrong, she can't find the correct answer t for step 4 and will cause Alice to reject in step 5, essentially. Again, the probability for Eve's choice to be correct is $1/S$.

Later: If Eve acts like a benign adversary in the first two steps in order to become active in the third or fourth step, it is infeasible for her to compute either a or the correct response values $r = h_1(a)$ or $t = h_3(a)$, essentially. Then her probability of success is negligible, i.e. $\ll 1/S$.

Hence, D3 holds for $R = 0$. Note that by excluding one candidate p_i from the set of possible candidates for p, we also exclude the corresponding weak secret π_i as a possible choice for π. At first there are S candidates $1, \ldots, S$, for π. After R rejections there remain $S - R$ such candidates, w.l.o.g. $1, \ldots, S - R$. By induction we verify that Eve's chance to be correct is $1/(S - R)$, then. $\square$

Note that Alice even may publish her public key E in advance (after the first execution of the protocol Eve knows E anyway) or use the very same public key to communicate with other participants than Bob, without risking her security.

4 RSA and protected OKE

In this section, we point out a problem with naive RSA-based generic OKE. It comes from Bob not being able to verify all our requirements for OKE to be satisfied. We propose protected OKE, a variant of generic OKE to address this problem.

At a first look, the RSA public-key cryptosystem appears to be ideal for open key exchange. The public key is the pair $E = (e, N)$, where $N = PQ$ is the product of two huge primes P and Q and e is relatively prime to $\varphi(N) = \varphi(P)\varphi(Q) = (P-1)(Q-1)$. The encryption function is defined by $E(x) = x^e \bmod N$, the decryption function is $D(x) = x^d \bmod N$, where the secret exponent d is chosen such that $ed \equiv 1 (\bmod \varphi(N))$. The natural choice for the group operation $\Diamond$ is the multiplication mod N, the group S_E is the set of all numbers in $\{1, \ldots, N-1\}$ which are not divisible by P or Q, i.e. the size of S_E is $|S_E| = \varphi(N)$.

In the case of RSA, all requirements are easy to verify—with one important exception: The function E must be invertible and hence the equation $D(E(x)) = x$ must hold.[6] In the case of RSA, this is no problem if e and N are chosen according to their specifications—but e and N can be chosen in a different way and Bob is unable to detect this.

We come up with some slightly changed requirements:

1. If chosen by an honest participant (i.e. Alice), the function $E : S_E \longrightarrow S_E$ is invertible.
2. If E is chosen by Eve, the space of the encryption function E can collapse.
3. As before, "$\Diamond$" is the group operation of S_E.
4. The finite group S_E is of order $|S_E| \geq 2^{2K}$.
5. $H' : \{0,1\}^* \longrightarrow S_E$ is a random function just like H.

All other requirements remain the same.[7]

Since "$\Diamond$" is a group operation, there is a number $k \in \{1, 2, \ldots, \}$, such that for every $x \in S_E$ the set $E(x)$ has k elements. If $k = 1$, the function E is invertible and we have no problem. Otherwise $k \geq 2$, hence for every ciphertext $y \in S_E$, $y = E(x)$ there are at least two possible decryptions x_1, and x_2, $y = E(x_1) = E(x_2)$.

This observation leads to the **protected OKE protocol**. For protected OKE, we we change generic OKE like this:

Instead of $\mu \in_R \{0,1\}^l$, Bob chooses 2 values μ_{-1}, $\mu_0 \in_R S_E$. Now Bob computes $\mu_1, \mu_2, \ldots, \mu_K$ by

$$\mu_i = E\left(\mu_{i-2} \Diamond H'\left(\mu_{i-1}\right)\right).$$

The last two values μ_{K-1} and μ_K are sent back, instead of μ in the case of the generic protocol.
The value p now is computed by $p = H(E|m|\mu_{-1}|\mu_0|\pi)..$

If E is invertible, the values μ_{-1} and μ_0 can be computed by K evaluations of the decryption function D and the random function H'.

---

[6] Otherwise, even if $a \in_R S_E$, the random variable $E(a)$ would not be distributed uniformly, hence Eve could use $q = E(a) \Diamond p$ to collect some knowledge about the actual value of p, especially to rule out some of the candidate values for p.

[7] In the case of RSA, all new requirements can easily be enforced by both Bob and Alice. For Bob, enforcing the third requirement means to verify $\gcd(x, N) = 1$ for the RSA-modulus N and all (pseudo-)random values $x \in_R \{1, \ldots, N-1\}$, $x = H(\ldots)$, or $x = H'(\ldots)$ he generates. For practical choices of the security parameters, the fourth requirement is satisfied, except when the RSA-modulus has lots of tiny prime factors. If the order of the group S_E were small, a random value $x \in_R \{1, \ldots, N-1\}$ would likely be $\notin S_E$. Since $x \notin S_E \Longleftrightarrow \gcd(x, N) \neq 1$, checking the gcd is sufficient to protect against violations of the third and fourth requirement.

If E is not invertible, given μ_{K-1} and μ_K there are at least two choices μ_{K-2}^1, μ_{K-2}^2 for μ_{K-2}, for every μ_{K-2}^i there are two choices for μ_{K-3}, ... and thus we expect 2^K choices $\mu_{-1}^1, \ldots, \mu_{-1}^{2^K}$ for μ_{-1}. Due to the randomness of H' and because $|S_E| \geq 2^{2K}$, the probability that there is a significant number of collisions $\mu_{-1}^i = \mu_{-1}^j$ for $i \neq j$ is negligible.[8] It is infeasible for Eve to compute all possible values for p, even if π is fixed. Thus, the partial knowledge about p she gets from inspecting $q = E(a)\Diamond p$ is useless for her.

This way, we have modified our protocol such that it is secure even if Bob can't directly check E to be invertible.

For a secure realization of protected RSA-OKE, the following choices are suggested: $K = 80$ for the security parameter K, $N \approx 2^{1000}$ for the modulus N and the use of a dedicated hash function like SHA-1 with 160 bits of output (or more) as the underlying primitive to realize pseudorandom functions H, H', h_1, h_2, and h_3. For the public encryption exponent e we may make a fixed choice which allows speedy RSA encryption, e.g. $e = 3$.

With these parameters, the communication costs are quite reasonable: Four 1000-bit numbers N, μ_{-1}, μ_0, and q and three 160-bit strings m, r, and t are exchanged between Alice and Bob.

5   A comparison between OKE and EKE protocols

In this section we compare EKE and OKE protocols, with a closer look on their variants based on RSA.

As its inventors write, *encrypted key exchange* (EKE) relies *"on the counter-intuitive notion of using a secret key to encrypt a public key"* [4], with the secret key being derived from the password. If π is a weak secret, known to Alice and Bob, $\pi(X)$ denotes the encryption of X using π. Similarily we write $k(X)$ for encrypting X using the session key k. For an asymmetric key pair (E, D) with the public key E, $E(X)$ denotes the public key encryption of X. The **generic EKE protocol** can be described like this:

1. At random, Alice generates an asymmetric key pair (E, D) and sends $\pi(E)$ to Bob.
2. Bob decrypts this to find the public key E, chooses a random session key k, and sends $\pi(E(k))$ to Alice.
3. Alice decrypts this to get k. Further, she chooses a random challenge c to send $k(c)$ to Bob.
4. Bob decrypts this, chooses a random challenge d, and sends $k(c|d)$ to Alice.
5. Alice checks whether Bobs reply is of the form $k(c|\ldots)$. If so, the sends $k(d)$ to Bob. Else, she rejects.
6. Bob checks whether Alice's last reply is correct and rejects if not.

Two properties of the EKE are

---

[8] This argument is based again on the birthday paradox.

(A) Every protocol execution requires Alice to generate a new key pair.[9]
(B) The EKE depends on encrypting the public key.

The impact of property (A) is that asymmetric cryptosystems with a slow key-generation process tend to be slow when running the protocol. While e.g. El-Gamal asymmetric encryption and Diffie-Hellman key exchange do not require a lengthy key generation process, cryptosystems like RSA do. RSA requires to find two huge primes. This is no serious problem for OKE protocols, since here, a key generated once can be used forever.

If one considers an RSA-based realization of EKE, things become quite difficult due to property (B). It is hard to see how some RSA-modulus N can be encrypted such that trial decryption using the wrong key is not likely to reveal the fact, that the result of the decryption is not a valid RSA modulus. The inventors of EKE, Bellovin and Merritt, suggest to send N in the clear and only to encrypt the public exponent e [4]. Here, e should be a randomly chosen odd integer. They point out how to choose two huge primes P and Q, such that odd random integers are relatively prime to $\varphi(PQ) = (P - 1)(Q - 1)$.

While the protected RSA-OKE allows us to choose the public exponent e as we like, e.g. $e = 3$ for fast RSA encryption, RSA-EKE requires e to be an odd random number, which slows down encryption. On the other hand, protected OKE requires Bob to encrypt many (i.e. $K + 1$) plaintexts with RSA and Alice to decrypt as much values.

The main advantage of protected RSA-OKE in comparison to RSA-EKE is that public keys are generated once and for all—RSA key generation is very demanding, especially for slow devices like smartcards. This is likely to overcome the main disadvantage of protected RSA-OKE: The need to compute $K + 1$ ciphertexts in the case of Bob and to decrypt all these ciphertexts in the case of Alice.

6 Related work

Apart from Bellovin and Merritt [4], only few other key-exchange and authentication protocols for weak common secrets have been designed.

Gong and others [7] presented protocols with the requirement that either authentic public keys had to be known, or had to be exchanged encrypted and

---

[9] Otherwise, there is a replay attack for Eve: First, she listens to a protocol execution where Alice sends the value $x = \pi(E)$ in the first round and Bob replies $y = \pi(E(k))$ in the second. Then she requests the session key k. If Alice uses $x = \pi(E)$ once again, Eve impersonates Bob and replies y. Now, the session key is k, which Eve knows.

Also, if Alice reuses E for communicating with another side, e.g. Dan, using a different password ρ, Eve can find both π and ρ by brute force, looking for two candidates π_i and ρ_j with equal decryptions $\pi_i^{-1}(x) = \rho_j^{-1}(z)$, where $x = \pi(E)$ and $z = \rho(E)$.

remain secret ("secret public key protocols"). For the second variant the authors expressed a concern about finding suitable asymmetric encryption methods. Gong [8] later optimized the protocols of [7]. Among his protocols was a three-round variant.

Another 3-round protocol is due to Steiner, Tsudik and Waidner [13], who optimized the Diffie-Hellman EKE protocol from Bellovin and Merritt.

Jablon [9] gave a very diligent treatment of the Diffie-Hellman EKE protocol and also presented SPEKE, a similar protocol without symmetric encryption.

Information leakage enables passive attackers to eliminate possible passwords. Patel [11] pointed out a possible source of information leakage in EKE protocols. Its effectiveness depends on the structure of the public keys of the asymmetric cryptosystem in use, the blocksize of the symmetric cryptosystem in use, and the padding rules applied when a public key is encrypted using the symmetric cryptosystem. Patel demonstrated that the "random padding" suggested by Bellovin and Merritt [4] can be dangerous.

An approach quite different from EKE-like protocols is "fortified key negotiation", due to Anderson and Lomas [1] (see also section 22.6 of [12]). After exchanging a session key using ordinary public key techniques, a special kind of hash function with many collisions is used to authenticate the session key. This hinders key-guessing attacks, but in general violates our requirement D3 because too much information about the password is leaked.

Bellare and Rogaway [3] presented a model for authenticated key exchange with *strong* common secrets which heavily influenced the model in this paper. At this workshop, Blake-Wilson and Menezes [5] also considered the model of Bellare and Rogaway.

7 Concluding remarks and open problem

For the sake of simplicity, the model in this paper is less general than the one of Bellare and Rogaway [3]. Apart from the fact that attackers can always run the trivial attack with significant probability of success, which is a direct consequence of using a weak common secret, all the restrictions of our model compared to the one of Bellare and Rogaway can be removed.

Simply spoken, a *secure mutual authentication protocol* is a two-party protocol where both parties accept (at least with overwhelming probability) if their adversary is benign, but reject with overwhelming probability in the presence of an active attacker. The formal definition of a secure mutual authentication protocol is lengthy and somewhat complicated, see [3] or Blake-Wilson's and Menezes' paper in these proceedings. Due to Bellare and Rogaway, *secure key exchange* protocols are secure mutual authentication protocols with additional (secure key-transport) properties. It is easy to see that our OKE protocol and its variant are secure mutual authentication protocols, except, of course, that an active attacker's probability of success does not significantly improve on the trivial attack—which is not negligible.

Also, Alice is always the initiator[10], Bob is always the responder, and there is only one initiator and one responder in our model. As was pointed out by Blake-Wilson [6], if the role of initiator and responder could change, Eve could run a replay attack with Alice being both the initiator and the responder. Without extra checks, Alice could accept two sessions seemingly with Bob, both sessions using the same session key. If Eve asked for the one of the session keys, she could successfully "guess" the other. By including the initiator's and the responder's names (or any unique identification strings) into the messages, we can get rid of this restriction, as did Bellare and Rogaway in their protocols.

Due to Jablon [9], *perfect forward secrecy* "means that disclosure of the password doesn't reveal prior recorded conversations". If Eve learns the password π, she can compute the values p of old conversations, but she can't find $a = D(q \neg p)$. On the other hand, disclosure of Alice's secret key E would allow Eve to compute $a_i = D(q \neg p_i)$ with $p_i = H(\ldots|\pi_i)$ for all possible secrets π, to find the value $a_i = a$ by comparing r and $h_1(a_i)$ and thus to decrypt old conversations. This might be a reason for Alice to change her asymmetric key pair (E, D) frequently.

One of our requirements was that the number S of choices for the password was small enough that brute-forcing S^2 values is still feasible. Actually, we made use of this in the context of equation (1). Does this mean that our protocol is insecure if our common secrets are too strong?

Of course not! In the proof we argue that if Eve can find more than one solution for equation (1), she can invert the encryption function E on one out of S^2 ciphertexts. Because this is infeasible for her, she only can exclude one out of S possible choices for the password π.

Now, if S is too large, we can't rule out that Eve can find $s > 1$ solutions for equation (1), exclude s possible choices for π and thus improve on the trivial attack. But this sort of information leakage is very limited, if s were too large, assumption A2 would no longer hold. In other words, as long as the password space is large, we may run the risk of some information leakage, but with decreasing password space the possible information leakage adaptively decreases.

EKE-like protocols seem to be well suited for using public key cryptosystems like Diffie-Hellman and ElGamal. Consequently, most of the recent work on EKE-like protocols [13, 9] concentrates on using Diffie-Hellman key exchange as the underlying public key cryptosystem.

In contrast to this, (protected) OKE protocols seem to be especially well suited for RSA. On the other hand, the ElGamal cryptosystem does not satisfy our requirements, i.e. encrypting with ElGamal can not be described by a function $E : S_E \longrightarrow S_E$. Thus the security proof is of no use here. But is ElGamal-OKE actually insecure?

Yes, it is! Jablon [10] found a simple attack on the generic OKE protocol if ElGamal encryption is used straightforwardly . The attack exploits the fact that ElGamal is a "disclosing encryption system", see end of section 2.4 of [4].

---

[10] The party which sends the first message in a two-party protocol is called the "initiator", the other party the "responder".

It is an open problem to find more efficient variants of (protected) OKE, e.g. a three round protocol, similar to the optimized protocols found in [8] and [13].

Acknowledgement

The author would like to thank Paul Syverson for providing some relevant references (especially [11]), Simon Blake-Wilson for an interesting discussion on the Bellare-Rogaway model, and David Jablon for many helpful comments on a preliminary version of this paper.

References

1. R. Anderson, M. Lomas, "Fortifying Key Negotiation Schemes with Poorly Chosen Passwords", *Electronics Letters*, Vol. 30, No. 13, 1994, 1040–1041.
2. M. Bellare, P. Rogaway, "Random Oracles are Practical: A Paradigm for Designing Efficient Protocols", *First ACM Conference on Computer and Communications Security*, ACM, 1993.
3. M. Bellare, P. Rogaway, "Entity Authentication and Key Distribution", *Crypto '93*, Springer LNCS 773.
4. S. Bellovin, M. Merritt, "Encrypted key exchange: Password-based protocols secure against dictionary attacks", *Proc. IEEE Computer Society Symposium on Research in Security and Privacy*, Oakland, 1992.[11]
5. S. Blake-Wilson, A. Menezes, "Security Proofs for Entity Authentication and Authenticated Key Transport Protocols Employing Asymmetric Techniques", 1997, these proceedings.
6. S. Blake-Wilson, private communication.
7. L. Gong, M. Lomas, R. Needham, J. Salzer, "Protecting Poorly Chosen Secrets from Guessing Attacks", *IEEE Journal on Selected Areas in Communications*, Vol. 11, No. 5, 1993, 648–656.
8. L. Gong, "Optimal Authentication Protocols Resistant to Password Guessing Attacks", *Proceedings of the 8th IEEE Computer Security Foundations Workshop*, 1995, 24–29.
9. D. Jablon, "Strong Password-Only Authenticated Key Exchange", *ACM Computer Communications Review*, October 1996.
10. D. Jablon, private communication.
11. S. Patel, "Information Leakage in Encrypted Key Exchange", manuscript.
12. B. Schneier "Applied Cryptography" (2nd ed.), Wiley, 1996.
13. M. Steiner, G. Tsudik, M. Waidner, "Refinement and Extension of Encrypted Key Exchange", *Operating Systems Review*, Vol. 29, No. 3, 22–30.

---

[11] The full version of this paper was found at
http://www-cse.ucsd.edu/users/mihir/papers/eakd.ps.gz.

Protocol Interactions and the Chosen Protocol Attack

John Kelsey   Bruce Schneier   David Wagner

Counterpane Systems
101 E. Minnehaha Parkway
Minneapolis, MN 55419
{kelsey,schneier}@counterpane.com

U.C. Berkeley
C.S. Div., Soda Hall
Berkeley, CA 94720-1776
daw@cs.berkeley.edu

Abstract. There are many cases in the literature in which reuse of the same key material for different functions can open up security holes. In this paper, we discuss such interactions between protocols, and present a new attack, called the chosen protocol attack, in which an attacker may write a new protocol using the same key material as a target protocol, which is individually very strong, but which interacts with the target protocol in a security-relevant way. We finish with a brief discussion of design principles to resist this class of attack.

1   Introduction

One of the most difficult engineering aspects of designing any secure system—cryptographic protocol, cryptographic primitive, etc.—is identifying all of the assumptions that may affect security. Most of the time, when designing cryptographic protocols, we silently assume that the only access anyone has to the keys involved is through the protocol steps, or other steps much like them.

However, the real world is not that clean: Tamper-resistant tokens can hold only a few public-key/private-key key pairs at one time; users are somtimes lucky to have even one certified public key. Given these sorts of design constraints, it's likely that the same private key(s) will be used for several different systems running on the same device. These systems will probably be designed by different people and fielded together without any analysis of their potential interactions.

A protocol may be quite secure alone, but may lose its security when another protocol exists that can be carried out with the same key pair. In fact, it is always possible, in principle, to defeat the security of protocol $\mathcal{P}$, if we are able to choose another protocol, $\mathcal{Q}$, to be run by the same participants in parallel with the target protocol. A general construction for protocol $\mathcal{Q}$ (the "chosen-protocol" attack) will be given below. In some cases, the chosen-protocol attack is not practical. However, on many systems such as smart cards and users' PCs, the attack can be both practical and effective.

In this paper, we discuss protocol interactions which can weaken the security of one or both protocols. We then describe a new attack, the "chosen protocol" attack, in which a new protocol is designed to interact with an existing protocol, to create a security hole. After discussing generalities, we give several specific examples of this attack. We then look at accidental protocol interactions. Finally, we discuss protocol design rules that appear to render the chosen protocol attack impossible.

2   Protocol Interactions

Let $\mathcal{P}$ and $\mathcal{Q}$ be two different protocols, both of which use the same key material, but which do different things. These protocols are said to **interact** whenever some information derived from $\mathcal{P}$ allows an attacker to successfully mount some attack on $\mathcal{Q}$. For example, suppose that both protocols rely for their replay-resistance on a digitally-signed timestamp from one party of the protocol. Then the protocols interact, since an attacker can now use his observations of the execution of $\mathcal{P}$ to mount a replay attack on $\mathcal{Q}$.

When $\mathcal{P} = \mathcal{Q}$, the possible protocol interactions reduce to a subset of possible attacks on the protocol. A replay attack (replaying messages from one instance of a protocol to attack another instance of the same protocol) is an example of an interaction between a protocol and itself, as is a standard man-in-the-middle attack.

There are a great many ways in which two different protocols can interact. For example, it may be possible for a third party to simply observe the messages in $\mathcal{P}$, and then mount an attack on $\mathcal{Q}$. Alternatively, it may be possible for an attacker to mount a variant of the man-in-the-middle attack: Alice believes she is executing protocol $\mathcal{P}$ with Bob, but instead, she is executing protocol $\mathcal{P}$ with Mallory, who uses the information derived to execute protocol $\mathcal{Q}$ with Bob in Alice's name. In still other cases, it may be possible for Mallory to execute protocol $\mathcal{P}$ in his own name with Alice, and simultaneously execute protocol $\mathcal{Q}$ with Bob in Alice's name. It is not necessary for the legitimate execution of $\mathcal{P}$ to complete successfully, so long as the attack on $\mathcal{Q}$ is made possible.

2.1   Sharing Keys Among Many Different Protocols

The one attribute common to all protocol interactions is that they involve the use of shared key material, either public/private key pairs or symmetric keys, between different protocols and applications. Most systems today use a different symmetric key or public/private key pair for each different function or application. This generally reduces the number of protocols having access to a given key or key pair to a manageable number. However, there are three forces pushing us toward a world in which different applications share common key material:

1. Certification. Certification of public keys is a costly process. Relatively few users are likely to want to maintain twenty certificates, paying a few dollars per year for each.
2. Cryptographic APIs. As cryptographic APIs become widespread, and more non-cryptographic applications make limited use of cryptography, it becomes

increasingly likely that many different protocols will emerge, all of which may make use of the user's certified public key by default.

3. Smart cards. Smart cards and other cryptographic tokens often have limits on the number of key pairs they can store. If many applications are allowed to use the same smart card, then there will likely be some use of keys for different applications.

Already VeriSign certificates are used to provide security for PEM [Lin93, Ken93, Bal93, Kal93, Sch95, Sch96], S/MIME [RSA96, Dus96], SSL [FKK96], and SET [VM96]. Entrust Technologies [Cur96, Oor96] markets technology to provide for a common key archetecture over multiple applications.

3 Accidental Interactions

There are several examples of very simple known-protocol attacks. (These attacks are generally much simpler than the chosen protocol attacks.)

3.1 RSA signing and encryption

It has long been known that carelessly providing RSA signatures with an RSA key used for encryption could lead to a simple attack: The attacker intercepts a public-key encrypted block intended for Alice, and then presents it to her, requesting a signature. If Alice complies, then the attacker ends up with the decryption.

3.2 RSA signing and zero knowledge

A standard way of doing zero-knowledge proofs using RSA also provides an RSA signing oracle for a clever attacker.[And97a, MOV97]

3.3 A Banking Protocol Interaction

In [AK97], Ross Anderson describes a potentially disasterous interaction between two mechanisms in a bank's system for managing ATM cards. The bank requested a program to update all its PIN numbers for a new systemwide key, and the hardware/software vendor provided such a program, with the warning that it should be run immediately and then deleted. This program could be used to recover the PIN numbers of anyone's card using a protocol interaction.

3.4 Blind and Regular Signatures

Using the same RSA modulus for both regular and blinded signatures can, naturally, allow various attacks. In most blind signature schemes, the signer does not know what he is signing; he might be signing anything. Therefore, the blind signer can be used as a RSA signing oracle to sign arbitrary messages, which can then be used to defraud the protocol using regular signatures.

3.5 Interactions Among Different Applications

One of the most interesting places we can get a protocol interaction is when an identical protocol is used for different purposes in different applications, and no signed or authenticated statements bind an execution of the protocol to its specific application.

One example threat is that a user may be authorized to execute one application on a secure server, but not another application. If both use the same protocol to establish the user's identity, then the user may be able to access functions from which he should be restricted. Another threat is that the user may voluntarily run one protocol with an outsider, who then can use this information to run another protocol with some other entity in the original user's name.

One fairly serious threat can come in the form of electronic mail packages that may be set up to automatically send a digitally-signed reply of a message. (For example, a person might be sent some large binary, with a note in the subject line saying "hit reply to be removed from our mailing list.")

4 The Chosen-Protocol Attack

The "chosen-protocol attack" is an attack in which some attacker convinces one or more intended victims to accept and start using a new, tailor-made protocol, called the "chosen protocol." This protocol is designed specifically to interact with some already-running protocol, called the "target protocol." The chosen protocol should have no obvious weaknesses, but must allow an attack on the target protocol.

A chosen protocol can always be built to interact the security of any given target protocol, if there are no restrictions on what the protocol steps are allowed. This can be shown by example. Suppose there is a target protocol which uses private key material only for signing and decrypting. If the chosen protocol gives the attacker a decryption oracle and a signing oracle, then the attacker is trivially able to defeat the target protocol. By installing an oracle of this type, we can trivially break any other protocol that shares the same key material. (For that matter, the first step of the chosen protocol could, in principle, simply be to send all the private key material to the attacker.)

It is more interesting to consider chosen protocols that aren't obviously dangerous or weak, but that still allow the attack on the target protocol. Below, we give a method for constructing chosen protocols which seems to satisfy this requirement. However, without a rigorous definition of what a reasonable-looking protocol is, it isn't possible to prove that reasonable-looking chosen protocols of this kind actually exist. One interesting definition of "reasonable-looking" is that the protocol is not susceptible to attacks based only on the messages that occur in this protocol. However, it is certainly not clear how to prove that any given protocol has this property.

We can give a more general argument for why chosen protocol attacks will generally exist for any target protocol. Here, we consider a protocol P and a protocol Q which are intended to be run between Alice and Bob. In the attack, Eve will run an instance of Q with Alice in Bob's name, and use information from this to run an instance of P with Bob, in Alice's name.

Consider a normal execution of P. Each time Alice needs to generate a protocol message, she must have the information necessary to do so. Thus, that information must be contained in messages she has received during P's execution, or information she already has. Each time Alice is supposed to receive some information from a protocol message, again, she must have the information to do so, and that information, once again, must come either from previous messages received during P, or from information Alice already has. Now, consider an instance of Eve executing P with Bob in Alice's name, and executing Q with Alice in Bob's name. The first protocol message of Q that Alice sends to Eve must contain the information needed to make the first message in P that is supposed to come from Alice. Whatever response Eve gets from Bob, the next couple of steps in Q can be used to recover that information from Alice and get it back to Eve, and to form the right response to Bob's messages. (Note that during these steps of Q, Eve can give Alice things to put into this response message, to accomplish her purposes.) This can continue for as long as is needed. Since Alice has all the information Eve needs, if she is willing to use that information as Eve wants it used, even without giving her a signing or decryption oracle, Alice will wind up allowing Eve to attack the other protocol.

This leads to the way to foil this attack: Make certain that Alice never gives Eve correct signatures, decryptions, MACs, etc., for any other protocol in the chosen protocol she runs with Eve. This will be discussed further below.

4.1 Justification

At first glance, the chosen protocol attack may look like a purely academic attack. However, we can provide some realistic scenarios for chosen-protocol attacks:

1. The same cryptographic keys may be reused by different products. For example, if the user's certified keypair is used both in a home-banking product and in a video game to confirm high scores, the lower-security product's protocols may be chosen to interact with the higher-security product's protocols.

2. Infiltration of lower-security products. We might expect a company designing an application to securely transmit financial or medical records over the internet to be very careful in the design of its protocols. However, the company may reuse the same key material in some lower-security protocols used in the same product. The design of these lower security protocols may not be as carefully overseen as that of the high-security protocols.

3. Protocols used in commercial products can often be strongly influenced by requests from important customers. Such requests could be used to install a chosen protocol for this class of attack. Such protocols may wind up being adopted later as worldwide standards.

5   Constructing Chosen Protocols

In this section, we present three chosen protocols which we believe to be plausible. Each is designed to work with the target protocol, so that sharing of keys makes some degree of sense. This is not a necessary trait of chosen protocols, although may provide plausable deniability to an attacker deliberately designing a chosen protocol.

5.1   Agora

In [GS96], the authors introduce Agora, a simple electronic payment protocol designed specifically for pay-per-view web pages. The protocol for making a payment can be described as follows. (In this description, Alice is buying something from Bob.)

Protocol A: Agora (the Target Protocol):

1. Alice (the customer) sends to Bob (the merchant):

   M_0 = Request for a price quotation from the merchant.

2. Bob forms:

   N = a running sequence number kept by the merchant,
   P = the price,
   $X_1 = MerchantCert_{Bob}, N, P,$

   and sends to Alice:

   $M_1 = X_1, SIGN_{SK_B}(X_1).$

   (He must also immediately increment N.)
3. If Alice wants to make the purchase, she verifies the certificate (including expiration date) and signature, forms:

   $X_2 = UserCert_{Alice}, N, P,$

   and sends to Bob

   $M_2 = X_2, SIGN_{SK_A}(X_2).$

4. Bob verifies Alice's certificate, her signature, and the values of N and P. If all is well, he delivers whatever was paid for. (In this context, this is probably a pay-per-view web page.)

Our aim is to build a special class of man-in-the-middle attack. In our attack, Alice thinks she's carrying out Protocol B (the chosen protocol) with Bob. Unfortunately for her, Mallory is sitting in the middle, using her protocol steps to allow him to carry out Protocol A (the target protocol) with Bob, in Alice's name.

For each step in the target protocol, we determine what information Mallory needs from Alice and from Bob in order to carry out the protocol. We can then design the chosen protocol to allow Mallory to get that information.

For example: In Step 3, Mallory needs

$$M_2 = X_2, SIGN_{SK_A}(UserCert_{Alice}, N, price)$$

from Alice. If he has this, he can successfully impersonate Alice to Bob. Therefor, Mallory also needs a plausible story behind a protocol designed to get him this information.

In light of concerns about children viewing inappropriate web pages, we might imagine another protocol, used with the same keys and certificates, for verification of adulthood with servers that don't charge for viewing their pages. We will design this chosen protocol so that it can be used by attacker Mallory to charge Alice for the web pages he views on Bob's page.

This age-verification protocol uses the same certificates as Agora, and works as follows:

Protocol B: Age-Verification Protocol (the Chosen Protocol):

1. Alice sends a request to Bob to view his page:

   $M_0 = $ Request to view page.

2. Bob responds by forming:

   $R_1 = $ A random challenge.
   $M_1 = R_1$.

   (Note that R_1 is designed to be the same size as the concatenation of the N and price in the Protocol A.)

3. Alice responds by forming:

   $X_2 = UserCert_A, R_1$.

   and sending

   $M_2 = X_2, SIGN_{SK_A}(X_2)$.

This protocol is secure when executed on its own, even though it has been designed specifically to subvert Protocol A.

To make this into a man-in-the-middle sort of attack, we put Mallory in Bob's place when Alice executes Protocol B (the age-verification protocol). Here is how Mallory uses his man-in-the-middle status with Alice in Protocol B to impersonate her in Protocol A:

1. Alice completex step 1 of Protocol B.
2. Mallory intercepts the message Alice sent to Bob in Step 1 of Proctocol B.
3. Mallory executes Step 1 of Protocol A with Bob.
4. Bob executes Step 2 of Protocol A with "Alice."
5. Mallory intercepts Bob's reply to Alice in Step 2 of Protocol A.
6. Mallory recovers N and P from M_1 of Protocol A and sends them to Alice as R_1, in Step 2 of Protocol B.
7. Alice responds with Step 3 of Protocol B.
8. Mallory allows to pass through to Bob.

At this point, Bob thinks he completed Protocol A with Alice, while Alice thinks she completed protocol B with Bob. Mallory is now free to intercept whatever information he forced Alice to buy from Bob.

This man-in-the-middle attack can work again and again. Each time Alice verifies her age to Mallory, he can now use her information to view web pages on an arbitrary merchants' pages, and stick her with the bill.

5.2 The Wide Mouth Frog Protocol

The Wide Mouth Frog protocol [BAN90, Sch96] is a well-known protocol for exchanging a symmetric encryption key using a trusted third party with whom each party shares a secret symmetric key. Its simplicity makes it an ideal example of how the chosen-protocol attack can work on symmetric, as well as public-key, protocols.

The key-exchange protocol, which will be our target protocol, works as follows[1]:

Protocol C: Wide Mouth Frog Protocol (the Target Protocol)

1. Alice wants to establish a session key with Bob. She begins by forming:

   K = a random 192-bit triple-DES key,
   T_A = current timestamp,

   and sends to Trent, the trusted third party,

   $M_0 = ID_A, E_{K_A}(T_A, ID_B, K)$.

2. Trent looks up the right secret key, K_A, and then decrypts the message and verifies the validity of the timestamp and ID_B. If all is well, he forms

   T_B = current timestamp (may be different than T_A),

   and sends to Bob

   $M_1 = E_{K_B}(T_B, ID_A, K)$.

3. At this point, Bob decrypts the message and verifies the timestamp and ID_A. If all is well, he now has a shared key with Alice, which he knows is authentic and fresh.

Our chosen protocol will be built to use this trusted third party and infrastructure of shared keys to allow secure logins. Now, we have a user, Alice, and a host, Mallory. This is how the basic protocol works:

Protocol D: Secure Login Protocol (the Chosen Protocol)

1. Mallory forms:

   R_0 = A 64-bit random number.
   T_x = Current Timestamp.

   and sends to Alice:

---

[1] We are filling in some specific values left open by the protocol's designers.

$$M_0 = LoginChallenge, T_x, R_0.$$

2. Alice responds by sending Trent:

$$M_1 = LoginRequest, ID_A, E_{K_A}(T_A, R_0, hash(passphrase), ID_M).$$

where T_A is the current timestamp, R_0 is the random number sent by Mallory, and ID_M and ID_A are Mallory's and Alice's IDs.

3. Trent verifies the timestamp and IDs, and then sends to Mallory:

$$M_2 = LoginMessage, E_{K_M}(T_M, hash(R_0, hash(passphrase)), ID_A).$$

4. At this point, Mallory verifies the timestamp and the hash. Note that Mallory only has access to the hash of the passphrase, and that outsiders never see even that.

Now, the chosen-protocol attack works as follows:

1. Mallory sends ID_B as R_0 in the first step of Protocol D.
2. Mallory catches Alice's response to Trent, strips away the *LoginRequest* header, and sends the rest of the message off to Trent as the first step of Protocol C.
3. Trent treats this as a valid request for a secure session with Bob from Alice. He sends the message to Bob.
4. Mallory now intercepts all messages from Bob to Alice, and impersonates her. Bob is convinced.

5.3   The DASS Public-Key Protocol

DASS (Distributed Authentication Security Service) is a protocol for mutual authentication and key exchange developed by Digital Equipment Corporation and marketed in a product called SPX [TAP90, TA91].

Protocol E: DASS (the Target Protocol)

– Protocol steps deleted for space considerations. See [Sch96].

To build a chosen protocol, we have to add some additional functionality to the system. In this example, we add a new protocol to have Trent generate random public keys for us, as needed.

Protocol F: Protocol for Requesting a Temporary Public Key (the Chosen Protocol)

1. Alice forms

$$R_0 = \text{a random number the same length as an ID,}$$

and sends to Trent

$$M_0 = RequestForPK, ID_A, R_0.$$

2. Trent generates a new public key, PK_T, forms

$$K_1 = \text{a random encryption key,}$$

and sends back

$$M_1 = PKE_{PK_A}(K_1), E_{K_1}(SK_T), SIGN_{SK_T}(R_0, PK_T).$$

With this, we can build a chosen-protocol attack based on being the man-in-the-middle between Bob and Trent. We choose R_0 to be some person's ID, and then use the signed block to convince Bob that it's the right public key. Thus, this protocol looks like:

1. Mallory selects

    $$R_0 = ID_A,$$

    and sends to Trent

    $$M_0 = RequestForPK, ID_M, R_0.$$

2. Trent generates a new public key, PK_T, forms

    $$K_1 = \text{a random encryption key,}$$

    and sends back to Mallory

    $$M_1 = PKE_{PK_M}(K_1), E_{K_1}(SK_T), SIGN_{SK_T}(R_0, PK_T).$$

3. Mallory now encrypts a random session key under Bob's public key, including the timestamp, key lifetime, and Alice's ID. All this is exactly as appears in the third message of DASS.
4. Bob sends ID_A to Trent.
5. Mallory intercepts this request. He sends back $SIGN_{SK_T}(R_0, PK_T)$, recovered from the second message in the chosen protocol.
6. Bob decrypts this message, and uses PK_T to verify the signature on the first message sent to Bob. Bob is now convinced he shares a secret symmetric key with Alice, when in fact, he shares it with Mallory instead.

6 Design Principles for Avoiding Weakening Interactions

To prevent protocol interactions, we impose a few requirements on all protocols implemented with a given key or key pair.[2] If these rules of thumb are followed, then we believe a chosen protocol attack cannot work.

1. The first and most important rule to follow is to limit the scope of each key. A key should typically have only a small number of closely related functions. There is sometimes a temptation to reuse keys for related applications—this should be avoided whenever possible. If there is only one certified key pair, it should be used to sign (and perhaps even derive) other single-use public keys, as in [And97b, SH97]. This eliminates the overwhelming majority of possible protocol interactions.

---

[2] Some of these rules were inspired by [And95].

2. Each application, protocol, version, and protocol step or operation that can be performed using a given key must have its own unique identifier, which must have a fixed length, and be used in a standard way in all protocol steps and operations. Note that it is very important that different versions of a protocol, or the same protocol running in two different applications, be differentiated here. The goal here is that each different use of a given key has a different unique identifier, and that this identifier is involved in whatever cryptographic operations are done using that key.

   This guideline is similar to the design approach, fail-stop protocols, advocated in [GS95]. There they suggested signing every protocol message, and including in each message a header containing the sender's name, the recipient's name, the protocol identifier and version number, a sequence number, and a nonce or timestamp. The point is to ensure that, if an active attacker injects a fake message, protocol execution will immediately halt. This allows one to consider only passive attacks when verifying the confidentiality of protocol secrets, which makes it possible to prove that secrets remain secret; this, in turn, allows one to apply BAN logic to verify the security of the protocol. In particular, their work on extensible fail-stop (and fail-safe) protocols provides significant progress towards ensuring composibility of protocols, even when private keys are reused. We conclude that the fail-stop design approach is valuable in the context of chosen-protocol attacks, and that including unique identifiers in each message is a powerful defensive design technique.

3. In message authentication and signing operations, simply including the fixed unique identifier a fixed place in the authentication operation will prevent chosen-protocol attacks from other messages that follow the same guidelines.

4. In public- and secret-key encryption operations and key-derivation operations, the unique identifier should be used in a way that makes the message impossible to decrypt without using the right unique identifier. There are several straightforward ways to do this: The simplest one conceptually is to use the protocol-identifier as a symmetric encryption key, and use it to encrypt whatever value is to be decrypted by the receiver in a way that an attacker cannot undo. For example:

   (a) When a public-key encryption is being used to send a symmetric encryption key, then that symmetric encryption key is encrypted under the protocol-identifier first, and then under the public key.

   (b) When symmetric encryption is being used, the message is first encrypted under the protocol-identifier, and then under the real symmetric encryption key. (Note that this will not work when the order of encryption is interchangeable, as in an OFB stream cipher.)

   (c) When symmetric encryption is being used that is order independent, or when the above guideline is too computationally expensive, the message is encrypted under a random symmetric key, and that key is encrypted and authenticated (in an order-dependent way, such as by using a block cipher) first with the protocol-identifier, and then with the actual secret symmetric key that would otherwise have encrypted the message.

(d) When a symmetric key is being derived from shared secret information (such as might result from a Diffie-Hellman key agreement operation), the symmetric key is derived by hashing together the shared secret information and the protocol identifier.

5. Smartcards should include support for, and enforcement of, these mechanisms. Smartcards should be aware of public key reuse across protocols and applications.

The basic guideline is that every time some key is used, there must be a cryptographic binding between the message produced and the unique identifier of that message. This binding ensures that a message in one protocol cannot be substituted for some message in another protocol adhering to the same guidelines. In authentication and signing operations, simply including the unique identifier in the authenticated or signed operation is sufficient to prevent inter-protocol reuse of the signed message. In encryption operations, trying to decrypt a message with the wrong protocol or step identifier simply gives us an incorrect decryption.

7 Conclusions

The chosen-protocol attack resonates with a fairly common theme in security: security does not necessarily compose. Two protocols can each be secure on their own, but when they are implemented together the composed system may no longer be secure.

The chosen protocol attack brings to mind Bob Morris's comments at Crypto '96; he discussed building systems that are secure "even when they contain a John Walker" [Mor96] or, more generally, a nameless insider intent on attacking or subverting the system. What do you do if you've got an attacker on the design team for one of the (many) cryptographic protocols you rely on? How do you compartmentalize for robust security? The Internet owes its success to its mass decentralization, and its clever uses of existing infrastructure components in new and useful ways; we need to ensure that Internet security can survive in this environment.

8 Acknowledgements

The authors wish to thank Ross Anderson, Susan Langford, Mark Lomas, James Riordan, Paul Syverson, and all those who made comments during and after the presentation at SPW'97 for their invaluable help in improving this paper.

References

And95.   R. Anderson, "Robustness Principles for Public Key Protocols," *Advances in Cryptology — CRYPTO '95*, Springer-Verlag, 1995, pp. 236-247.

And97a.   R. Anderson personal communication, 1997.

And97b.   R. Anderson, "Perfect Forward Secrecy", presented at the rump session of Eurocrypt '97, 1997.

AK97.   R. Anderson, M. Kuhn, "Low Cost Attacks on Tamper Resistant Devices," these proceedings.

Bal93.   D. Balenson, "Privacy Enhancement for Internet Electronic Mail: Part III— Algorithms, Modes, and Identifiers," RFC 1423, Feb 1993.

BAN90.   M. Burrows, M. Abadi, and R. Needham, "A Logic of Authentication," *ACM Transactions on Computer Systems*, v. 8, n. 1, Feb 1990, pp. 18–36.

Cur96.   I. Curry, "Entrust Overview, Version 1.0," Entrust Technologies, Oct. 96. http://www.entrust.com/downloads/overview.pdf

Dus96.   S. Dusse, "S/MIME Message Specification: PKCS Security Services for MIME," IETF Networking Group Internet Draft, Sep 1996. ftp://ietf.org/internet-drafts/draft-dusse-mime-msg-spec-00.txt

FKK96.   A. Freier, P. Karlton, and P. Kocher, "The SSL Protocol Version 3.0", ftp://ftp.netscape.com/pub/review/ssl-spec.tar.Z, March 4 1996, Internet Draft, work in progress.

GS95.   L. Gong and P. Syverson, "Fail-Stop Protocols: An Approach to Designing Secure Protocols," *Fifth International Working Conference on Dependable Computing for Critical Applications*, Sept. 1995.

GS96.   E. Gabber and A. Silberschatz, "Agora: A Minimal Distributed Protocol for Electronic Commerce," *The Second USENIX Workshop on Electronic Commerce Proceedings*, USENIX Association, 1996, pp. 223–232.

Kal93.   B.S. Kaliski, "Privacy Enhancement for Internet Electronic Mail: Part IV— Key Certificates and Related Services," RFC 1424, Feb 1993.

Ken93.   S.T Kent, "Privacy Enhancement for Internet Electronic Mail: Part II— Certificate Based Key Management," RFC 1422, Feb 1993.

Lin93.   J. Linn, "Privacy Enhancement for Internet Electronic Mail: Part I— Message Encipherment and Authentication Procedures," RFC 1421, Feb 1993.

Mor96.   R. Morris, invited talk at Crypto '96.

MOV97.   A. J. Menezes, P. C. Van Oorschot, S. A. Vanstone, *Handbook of Applied Cryptography*, p. 418, CRC Press, 1997.

Oor96.   P.C.   van   Ooorschot, "Standards Supported by Entrust, Version 2.0," Entrust Technologies, Dec 1996. http://www.entrust.com/downloads/standards.pdf

RSA96.   RSA Data Security, Inc., "S/MIME Implementation Guide Interoperability Profiles, Version 2," S/MIME Editor, Draft, Oct 1996. ftp://ftp.rsa.com/pub/S-MIME/IMPGV2.txt

Sch96.   B. Schneier, *Applied Cryptography, Second Edition*, John Wiley & Sons, 1996.

Sch95.   B. Schneier, *E-Mail Security*, John Wiley & Sons, 1995.

SH97.   B. Schneier and C. Hall, "An Improved E-Mail Security Protocol," in preparation.

TA91.   J. Tardo and K. Alagappan, "SPX: Global Authentication Using Public Key Certificates," *Proceedings of the 1991 IEEE Computer Society Symposium on Security and Privacy*, 1991, pp. 232–244.

TAP90.   J. Tardo, K. Alagappan, and R. Pitkin, "Public Key Based Authentication Using Internet Certificates," *USENIX Security II Workshop Proceedings*, 1990, pp. 121–123.

VM96.   Visa and MasterCard, "Secure Electronic Transaction (SET) Specification, Books
         1-3" June 1996, http://www.visa.com.cgi-bin/vee/sf/set/intro.html
         or http://www.mastercard.com/set/set.htm.

Binding Bit Patterns to Real World Entities.

Bruce Christianson
James A. Malcolm

Faculty of Engineering and Information Sciences
University of Hertfordshire : Hatfield
England, Europe

Abstract. Public key cryptography is often used to verify the integrity
of a piece of data, or more generally to ensure that operations which
modify the data have been requested and carried out by principals who
are authorized to do so. This requires keys to be bound to principals in
an unforgeably verifiable manner.

Cryptographic bit patterns such as electronic key certificates (EKCs)
have a part to play in establishing such bindings, but the requirement
ultimately to bind keys to real world entities imposes subtle constraints
upon the structure and semantics of EKCs and related entities such as
ACLs and capabilities, and upon the role which such entities may play
in access control and integrity verification. These do not appear to be
adequately realized at present.

1 Introduction

In this position paper, our primary motivation is the problem of ensuring integrity in a wide open system. Enforcement of integrity has two aspects: firstly ensuring that any action which updates the system state is properly authorized (ie ensuring that the right thing is done) and secondly ensuring that the corresponding transaction is properly carried out (ie ensuring that the thing is done right.)

By a wide-open system we mean that network and operating system resources may be shared with strangers, or with other principals whom we do not trust. We also mean that integrity threats may arise from system "insiders" such as our colleagues and system managers, who should therefore be trusted only explicitly, and as little as possible. We do not assume that security domains (ie resources subject to a particular underlying integrity semantics) map neatly onto administration domains: indeed we assume that on some systems resources administered by the same authority may need to be be shared between security domains. (An example is a joint software development project which is contractually subject to certain QA procedures but where one of the subcontractors is also in competition with a different part of the prime contractor on another project.)

In order to check authorization for integrity purposes we frequently need to establish unambiguous verifiable bindings between a bit pattern (such as a name, a cryptographic key, or a program text) and a real-world entity (such as a person, a smart card, or a process running on a particular machine.) The difference between these two types of entity is that entities of the first type are bit patterns, which can be freely copied and modified in cyberspace; whereas entities of the second type have provenance in the physical world.

Our point of departure is the observation that real world artifacts cannot be authenticated remotely, except by binding them to bit patterns first. Although cryptographic techniques such as digital signature can bind bit patterns to other bit patterns, bit patterns alone cannot suffice to bind bit patterns to unique real world artifacts. A bridge between the electronic world of bit patterns and the "real" world of physical existence must itself have physical provenance as well as containing a representation of the bit pattern. Any attempt to verify the binding between a bit pattern and a real world entity using only electronic certificates is therefore doomed.

2   Binding keys to principals

A public key is a pattern of bits. A principal may be a human, or a piece of software running on a particular system, but in any event a principal is a real-world entity, with properties such as physical location. It is to this real world entity that a key must (eventually) be bound. A binding between a bit pattern and a real world entity cannot be effected solely by another bit pattern such as an Electronic Key Certificate (EKC).

The most an EKC can do is to bind a key to another bit pattern, such as a name or a program text. This is not sufficient to establish the identity (authenticity) of the corresponding real-world entity, since many such entities may "correspond" to the same bit pattern. Bit patterns can be copied, and neither is then the "original". But the binding must indicate which correspondence is intended.

It follows that some real-world artifact is required in order to complete the binding. This could be a copy of the EKC which resides in a particular physically protected location or address space, together with a social context in which the semantics of the protected copy are unambiguous. For example, a legally notarized paper copy of the EKC with established provenance stored in a bank vault would usually do. But the integrity and authentication of electronic access to remote real world entities is problematic, because cryptography protects only bit patterns.

Consequently a primary objective of an access and integrity control system is to reduce to a manageable level the number of times that physical access to a remote real-world entity is actually required in order to resolve a dispute.

One way of doing this is to provide electronic mechanisms which convince honest parties that they are in the right and which convince dishonest parties that recourse to the physical artifacts would in fact resolve the dispute against them.

3 Binding keys to access rights

Access rights must also (at some point) be bound to principals. For our purpose here, we can think of an access right as the capability to invoke a particular transaction (eg to login through a firewall). An access control mechanism based upon signature verification must answer the question: will signature with a key K suffice to satisfy an auditor that a request to invoke transaction T was valid?

It is tempting to bind keys and access rights to principals thus:

(bad)   key $\longrightarrow$ principal $\longleftarrow$ capability

The reason this is a bad idea is that it connects two bit patterns via a real world artifact, and thus makes both halves of the link from key to capability impossible to verify electronically. In practice, the principal is usually replaced for verification purposes by a bit pattern which represents its name:

(worse)   key $\longrightarrow$ name $\longleftarrow$ capability

In a wide open system, we cannot ensure that such a bit pattern is unique, and therefore the name alone does not suffice to ensure that both key and access right were intended to bind to the same (real world) principal. A better way of doing things is:

(better)   principal $\longrightarrow$ key $\longrightarrow$ capability

where of course either binding may be direct or indirect. This is better because public keys are unique by hypothesis: the inability of two parties to generate the same private key independently (whether by accident or design) is part of the foundation of public key cryptography. Of course, there is a chance that A could guess B's key by tossing a lucky coin repeatedly, but this probability can (and should) be made smaller than the chance of A and B being simultaneously struck by a meteor.

4 Indirect bindings

Indirect bindings can be used to reduce the amount of actual change required by updates to access control permissions. One common example of indirect binding is where keys are bound to rights via roles: one table indicates which keys may act in a particular role, and a second table indicates which roles may invoke a particular transaction:

principal $\longrightarrow$ key $\longrightarrow$ role $\longrightarrow$ transaction

The bit pattern representing a role has a similar uniqueness requirement to that for a key. One way of achieving this is to bind a public key of the role manager into the role name.

A similar approach can be adopted to the other half of the verification problem: how should a particular transaction be carried out correctly, and was it? A transaction is typically made up of a number of micro-operations (or method invocations) such as read, overwrite and append, which must be carried out on specific resources for the transaction to be valid:

principal $\longrightarrow$ key $\longrightarrow$ role $\longrightarrow$ transaction $\longleftarrow$ method $\longleftarrow$ key $\longleftarrow$ system

Further tables must specify these bindings. To allow for hardware maintenance there may be a further level of indirection mapping virtual resources to physical systems. This process can clearly be continued: the point to note is that the tables (or relation components) representing the inner links of the chains can be treated purely as bit patterns for the purpose of access control, and therefore can be cryptographically protected. Mappings from bit patterns to real-world entities should occur only at the boundaries of cyberspace, not in the middle of the chain.

5 Binding keys to hardware

Internally to the electronic world, real-world principals and systems are represented by keys. How should the bindings at the boundary be effected? A signature is generated in practice by applying a copy of a private key within a particular piece of hardware. Since moving private keys from one piece of hardware to another is a serious security risk, it follows that each key should be permanently bound to a single, identifiable piece of hardware, and physical (rather than cryptographic) means should be used to bind the hardware to specific places or people. (In general, we will need to bind several keys to the same piece of hardware in order to prevent protocol attacks.)

Since principals will use different pieces of hardware from time to time, it follows that a single principal will in general have many keys, which they use to exercise different rights. These rights must be explicitly delegated from key to key, using unforgeable instruments which can be recorded in an audit trail: from the key representing the owner of the resource to the key representing the principal authorized invoke some operation and thence to the key which will certify that the requested operation has been correctly carried out.

It must be possible for such delegation relationships to be verified remotely, and it therefore follows that delegation relationships must be expressed as relationships between bit patterns (keys).

Recent improvements in distributed systems architecture and infrastructure were originally motivated by the desire to allow more freedom over which piece of hardware performed the operations. From the point of view of integrity, however, transparency is a very bad thing, particularly when it extends to the point where examination of an audit trail cannot determine, even in principle, upon which piece of hardware a particular operation took place.

Fortunately, the same mechanisms developed to facilitate transparency can also be used in reverse: the requirement that certain operations be performed only upon particular pieces of hardware where particular keys are located is ceasing to be restrictive.

Relatively rigid bindings between keys and physical objects or locations can be exploited to allow a greater degree of assurance that the operations have been properly authorized and properly carried out, for example by ensuring that malfeasance would require active cooperation between mutually mistrusting administrations.

6 Binding principals to biometric data

Various mechanisms can be used to bind a bit pattern (actually a family of different "acceptable" bit patterns) to a physical person. These mechanisms may use retinal patterns or other vein structures which are only present in living people, and at most one such. These mechanisms are extremely useful, but do nothing to circumvent the need to bind access rights to principals only via cryptographic keys held in trusted physically secure or tamper-evident hardware.

Establishing a binding between a person and their biometric data in the first instance requires physical provenance of the person being certified, except in those cases where we care only that the same person be used on separate occasions, and do not care who they actually are.

Subsequent verification of the match between person and bit pattern requires that trusted hardware be in the vicinity of the person. Remote verification of identity in a wide open system additionally requires that the trusted hardware contain a cryptographic key. So we are again in the familiar word of principals being represented by keys which are associated with hardware which is in some sense trusted.

Adding the biometric data to the audit trail proves nothing: a replay is always possible, since the biometric data itself cannot be kept secret except by hardware which is trusted and tamperproof to begin with. Indeed, if biodata are used instead of keys then separating the different roles used by the same physical person requires an even greater degree of trust in the hardware.

The biometric hardware could be removed from the trust envelope if it were possible to use a living physical person in the same way as an uncopyable secret key. This would require using the person to produce a bit pattern which responds to a challenge and which can be verified without the person present, but which cannot be produced without them. This is a formidable research problem.

In the absence of such an advance, the use of biometric data can at most provide a cheap logical equivalent of a locked room with an armed guard on the door. If A and B are willing to believe the integrity of the hardware used to verify A's biometric data, then A will be willing to allow signature by the key associated with that hardware as a substitute for evidence of his physical presence for the purpose of audit, and B will be willing to delegate rights to the key directly. This then allows all remote verification by third parties (including audit mechanisms) to take place in the electronic (bit pattern) world. Delegation to remote hardware is expressed as a relationship between bit patterns.

7 Minimizing trust

Clearly a principal should not be empowered to act upon a resource unless the resource owner (or their delegatee) consents. However in a wide open system, the consent of the resource user is equally important: it should not be possible to confer a right unilaterally.

The principle of least privilege says that users should not be given rights they don't need. The obverse principle is that users should not be given rights they don't want. This is the principle of greatest consent : it should not be possible for a principal to acquire a right which that principal has not explicitly agreed to have. The purpose of enforcing this principle is to reduce the need for principals to trust remote systems, by making it harder for a principal to be framed (falsely accused of wrongdoing.) Note that the principle of consent protects the user who doesn't want to trust their system, as well as the server who doesn't want to trust their client.

For example, imagine that B's smart card (reported stolen) is discovered to contain a capability allowing a massive cashpoint withdrawal from an account belonging to the noted public figure A. A claims that B is holding him to ransom. B claims that A has stolen her card and inserted the capability himself. His objective is to blackmail her, using the ransom argument as the threat.

Clearly it is to the advantage of the honest party (whichever they are) that it should be impossible for anybody to place A's capability upon B's card without her prior knowledge and consent.

The principle of consent, combined with the principle of binding keys to hardware, leads to certain consequences: any access right which is bound to a key must be explicitly accepted by the corresponding principal, but the key used to accept a right need not be the key used to exercise the right.

For example, if R is a right, then the right to delegate R is another right R', and the right to accept R is another right R''. The rights R' and R'' may also be delegated from keys to other keys.

Similar remarks apply to the delegation of responsibility: the consent of both parties is required, but the keys which certify this consent may have a very obscure relation ship with the principals.

The verification of the web of delegation should not rely upon knowing who the principals involved are in real life. All interior bindings should be in cyberspace, and in any application where revocation is an issue, there will usually need to be lots of different keys (corresponding to physical places) where the delegation chain can be broken. Once again we see the importance of ensuring that delegation relationships can be verified in the world of bit patterns, without reliance upon knowing the bindings to real-world principals.

8 Auditing

The web of delegation must be recorded and verified by various parties, and this requires an explicit representation of the delegation policy, with guaranteed integrity. Operations must also be properly authorized, and this requires verifiably correct management and distribution of the tables which govern the interior bindings between keys and other bit patterns, and which hence confer access rights. It is important to assure integrity and access control properties for tables such as ACLs, transaction definitions, and role definitions.

Since we have reduced the tables to bit patterns, we can use cryptographic mechanisms to protect them as they move about the system. Access rights to tables are themselves capabilities. Proxies, capabilities, delegation certificates etc can be thought of as small pieces of a (virtual) table in transit. Updates to tables can be treated in the same way as other transactions. For example, unbinding a key from a role effectively revokes that key.

The access control system must be able to convince an auditor that the correct versions of the tables were used to control a given access and that only properly authorized updates to the tables have been made between versions. The audit trail should also enable the parties involved in a series of transactions to determine and undo the damage if the wrong versions were used, and to decide who should bear the cost of this.

Re-entrant application of the same basic access and version control mechanisms used for other shared resources such as files, combined with Optimistic Concurrency Control techniques for recovery, provide a good start for such an integrity mechanism, but the relations between keys and other bitstrings representing roles and rights are subtle.

These relationships need to be encapsulated, passed around a wide open system, verified and audited. It must be possible to make access control decisions which can be justified to an audit mechanism without requiring a prior physical check of the bindings between keys and real world entities in remote parts of the system.

The format and semantics of the next generation of electronic certificates need to be such that end-to-end guarantees can be made which provide all parties with an acceptable basis for proceeding, without covertly requiring that EKCs without secure real-world provenance be treated as having the same status as physical evidence.

The representation and manipulation of delegation relationships will play an important part in achieving this objective. The database community do not attempt to represent real world relationships other than via keys, some of which correspond to real world entities and some of which refer to further entries (relationships) in the database. An analogous approach using cryptographic keys may prove fruitful in the quest for integrity in wide open systems.

9 Conclusions

Real world artifacts have provenance. Real world artifacts cannot be authenticated remotely, except by binding them to bit patterns first. A primary objective of an access and integrity control system is to reduce to a manageable level the number of times that physical access to a remote real-world entity is actually required in order to resolve a dispute.

Bit patterns don't have provenance. Cryptographic techniques such as digital signature can bind bit patterns to other bit patterns, and such bindings can be verified remotely. A binding between a real world entity and a bit pattern cannot be effected solely by other bit patterns, but requires some real world artifact such as a piece of paper or hardware. Any attempt to verify the binding between a bit pattern and a real world entity using only electronic certificates is therefore doomed.

Rights should be bound to keys not to principals. Mappings from bit patterns to real-world entities should occur only at the boundaries of cyberspace, not in the middle of it. Bit patterns should be bound to each other via unique bit patterns and not via real world entities. The only unique bit patterns are those derived from private keys. Trusted hardware is still required.

Keys should be bound to hardware (not to names) and physical means should be used to manage the hardware and to effect the bindings on the boundary of cyberspace. Moving keys from one piece of hardware to another is a serious security risk. Don't do it. Biometrics doesn't reduce the trust requirements for the hardware needed to solve the binding problem, although it does provide a cheap locked room.

Transparency is bad for you, but the good news is that the requirement to do certain things only in certain places is ceasing to be restrictive. Relatively rigid bindings between keys and physical objects or locations can be exploited to allow a greater degree of assurance that the operations have been properly authorized and properly carried out.

Explicit delegation is required both to represent the security policy and to provide an audit trail. Rights may be delegated from keys to other keys. Delegated rights must be explicitly accepted. The key used to exercise the right may not be the key used to accept it. Delegated rights may include the right to accept or to confer other rights, as well as the right to exercise a right or to delegate further the right to exercise it.

Delegation should be expressed as a relationship between consenting bit patterns. The verification of the web of delegation should not rely upon knowing who the principals involved are in real life. If revocation is to be supported, it must be easy to break the delegation chain in lots of places.

An EKC is best thought of as a little piece of an ACL in transit. Electronic certificates cannot be adequate substitutes for the paper versions, and should not try to be. Both are required: their purpose is different.

References

1. Ellison, C.M., 1996, Establishing Identity without Certification Authorities, Sixth USENIX Security Symposium Procedings 67–76
2. Ellison, C.M., Frantz, B. and Thomas, B.M., 1996, Simple Public Key Certificate, http://www.clark.net/pub/cme/
3. Harbison, W.S., 1997, Trusting in Computer Systems, PhD thesis, Computer Laboratory, University of Cambridge
4. Lek, H. van der, Bakema, G.P. and Zwart, J.P.C., 1992, De Unificatie van Objecttypen en Feittypen een Pracktisch en Diadactisch Vruchtbare Theorie (Unifying Object Types and Fact Types : A Practically and Didactically Productive Theory), Informatie **34**(5) 279–295
5. Low, M.R. and Christianson, B., 1994, Self Authenticating Proxies, The Computer Journal **37**(5) 422–428
6. Needham, R., 1997, The changing Environment for Security Protocols, IEEE Network **11**(3) 12–15
7. Rivest, R.L. and Lampson, B., 1996, SDSI - A Simple Distributed Security Infrastructure, http://theory.lcs.mit.edu/~rivest/
8. Snook, J.F., 1992, Towards Secure Optimistic Distributed Open Systems, PhD thesis, University of Hertfordshire : Hatfield
9. Roe, M., 1997, Cryptography and Evidence, PhD thesis, Computer Laboratory, University of Cambridge

Breaking Public Key Cryptosystems on Tamper Resistant Devices in the Presence of Transient Faults

F. Bao, R. H. Deng, Y. Han, A. Jeng, A. D. Narasimhalu, T. Ngair

Institute of Systems Science
National University of Singapore
{baofeng, deng, yfhan, jeng, desai, teowhin}@iss.nus.sg

Abstract. In this paper we present a method of attacking public-key cryptosystems (PKCs) on tamper resistant devices. The attack makes use of transient faults and seems applicable to many types of PKCs. In particular, we show how to attack the RSA, the ElGamal signature scheme, the Schnorr signature scheme, and the DSA. We also present some possible methods to counter the attack.

1 Introduction

In September 1996, Boneh, DeMillo and Lipton from Bellcore announced a new type of cryptanalytic attack against RSA-like public key cryptosystems on tamper resistant devices such as smart card [4]. However, technical details of the Bellcore attack were withheld in that announcement and was released only at the end of October 1996. On 18th October 1996, Biham and Shamir published their attack, called Differential Fault Analysis (DFA), to secret key cryptosystems [5], such as DES. Some concrete ideas on how their attack works were revealed in their announcement.

Our work here was motivated first by the Bellcore announcement and then by the DFA announcement. Our first report on attacking RSA and some countermeasures were posted in the Internet on the 23rd and 24th October 1996 [2]. Right after that, A. K. Lenstra sent us his memo [9] on attacking RSA in Chinese remainder in a private communication. Subsequently, we released a more complete research note on attacking RSA and the ElGamal signature scheme on the 29th October 1996 [3]. Recently, Joye and Quisquater extended the Chinese remaindering attack to LUC and Demytko cryptosystems [8].

In this paper, we continue our earlier effort of attacking public-key cryptosystems (PKCs) on tamper resistant devices. Our attacking model makes use of the transient faults and seems applicable to many types of PKCs, such as RSA-like schemes and discrete logarithm based schemes. As in the Bellcore and DFA announcements, we assume that by exposing a sealed tamper resistant device such as a smart card to certain physical effects (e.g., ionizing or microwave radiation), one can induce with reasonable probability faults at random bit locations in a tamper resistant device at some random intermediate stage in the cryptographic computation. The faults in the random bit locations do not influence the code itself, i.e., the program itself does not crash, and only some of the values it operates upon are affected. It is further assumed that the attacker is in physical possession of the tamper resistant device and that he can repeat the experiment with the same private key by applying external physical effects to obtain faulty outputs.

The organization of the paper is as follows. In Section 2, we first report our attacks to RSA, and then present Lenstra's attack to RSA implemented based on the Chinese Remainder Algorithm (CRA). In Section 3, we show how to break discrete logarithm based schemes such as the ElGamal signature scheme [7], the Schnorr signature scheme [11], and the Digital Signature Algorithm (DSA). At the end of each section, we also give some possible methods to counter the attack.

When this manuscript was near its completion, Dan Boneh kindly sent us their paper [6] in a private communication. Throughout this paper, we will make remarks about the relation between their paper and our work wherever it is appropriate.

2 Attacking the RSA Scheme

Let $n = pq$ be the product of two primes p and q in RSA, e be the public exponent which is publicly known and d be the private exponent which is stored inside the tamper resistant device. Our attacks to RSA will be described in terms of ciphertext decryption although they can also be described in terms of signature generation.

Let m be a plaintext, then the corresponding ciphertext is

$$c \equiv m^e \bmod n$$

Denote the binary representation of the private exponent as $d_{t-1}|d_{t-2}| \cdots |d_i|$ $\cdots |d_1|d_0$, where d_i, taking value 1 or 0, is the ith bit, t is the number of bits in d, and $x|y$ denotes concatenation of x and y. Further, we denote

$$c_i \equiv c^{2^i} \bmod n, \text{ for } i = 0, 1, 2, ..., t - 1$$

Given c and d, the corresponding plaintext m can be expressed as

$$m \equiv c^d \bmod n \equiv c_{t-1}^{d_{t-1}} \cdots c_i^{d_i} \cdots c_1^{d_1} c_0^{d_0} \bmod n$$

2.1 Attack I

For the sake of simplicity, here we assume that in decrypting a ciphertext a single bit error is induced in c_i, for a random $i \in \{0, 1, 2, ..., t - 1\}$. Denote the corrupted value as c_i'. Then the output from the tamper resistant device is

$$m' \equiv c_{t-1}^{d_{t-1}} \cdots c_i'^{d_i} \cdots c_1^{d_1} c_0^{d_0} \bmod n$$

The attacker now has both m and m' so that he is able to compute

$$\frac{m'}{m} \equiv \frac{c_i'^{d_i}}{c_i^{d_i}} \bmod n$$

which equals $\frac{c_i'}{c_i} \bmod n$ if $d_i = 1$ or equals 1 if $d_i = 0$. (From now on we assume that every number we meet is relatively prime with respect to n, hence we can compute its inverse.) The attacker can easily compute all the possible $\frac{c_i'}{c_i} \bmod n$ values in advance (there are a total of t^2 such values since c_i' has t possible values). Now the attacker compares all these values with $\frac{m'}{m} \bmod n$. Once a match is found, he knows i and then knows that d_i is 1.

This simple example is just meant to illustrate the basic ideas of our attack. It showed that one bit fault at certain location and time can cause fatal leakage of the private key.

The example above assumes that only one c_i contains a single bit error and that there is no error propagation from c_i to c_j, $j > i$. The effects of such error propagation were considered in [6] and [9]. As a result, the error models in [6] and [9] are more complicated and probably more realistic than ours. From practical viewpoint, our model can be explained as the model for "read" error. That is, c_i is mistaken as c_i' when it is multiplied to the value for computing c^d but remains correct when it is squared to obtain c_{i+1}.

Another issue is that we can actually consider multi-bit faults instead of one bit fault only. In this case, we need to compare $m'/m \bmod n$ with many more possible values. For the case of two-bit faults, $m'/m \bmod n$ should be matched with all the values $c_{i_1}' c_{i_2}'/c_{i_1} c_{i_2} \bmod n$ ($i_1, i_2 \in \{0, 1, 2, ..., t-1\}$) and $c_i''/c_i \bmod n$, where c_i'' denotes the value of two bit errors in c_i. In this case, $O(t^4)$ possible values should be generated in advance and matched (as well as those $c_i'/c_i \bmod n$) with the value $m'/m \bmod n$. In general, about t^{2j} values need to be generated in the situation where j-bit faults may take place.

2.2 Attack II

Suppose that one bit in the binary representation of d is flipped and that the faulty bit position is randomly located. An attacker arbitrarily chooses a plaintext m and computes the ciphertext c. He then asks the tamper resistant device to decrypt c and induces a random bit error in d by applying external physical effects to the device. Assuming that d_i is changed to its complement d_i', then the output of the device will be

$$m' \equiv c_{t-1}^{d_{t-1}} \cdots c_i^{d_i'} \cdots c_1^{d_1} c_0^{d_0} \bmod n$$

Since the attacker now possesses both m and m', he can compute

$$\frac{m'}{m} \equiv \frac{c_i^{d_i'}}{c_i^{d_i}} \bmod n.$$

Obviously, if $m'/m \equiv 1/c_i \bmod n$, then $d_i = 1$, and if $m'/m = c_i \bmod n$, then $d_i = 0$. Therefore, the attacker can compare $m'/m \bmod n$ to $c_i \bmod n$ and $c_i^{-1} \bmod n$, for $i = 0, 1, ..., t-1$, in order to determine one bit of d. He repeats the above process using either the same plaintext/ciphertext pair or using different plaintext/ciphertext pairs until enough information in d is obtained.

Suppose one bit error takes place randomly in d in each fault test. Then by basic probabilistic counting, we have the following: If we take $t \log t$ fault tests, with a probability larger than half, every bit of d is disclosed.

It should be noted again that this attack applies to the case of multiple bit errors. Assuming two bit faults. The attacker needs to compare $m'/m \bmod n$ with $c_i c_j \bmod n$, $c_i/c_j \bmod n$, and $1/(c_i c_j) \bmod n$, for all $i, j \in \{0, 1, 2, ..., t-1\}$. In this case, matching $m'/m \bmod n$ with all these values has a complexity of $O(t^2)$ instead of $O(t)$ as in the single error case; while with large possibility one obtain two bits, d_i and d_j, once a successful match is obtained.

2.3 Lenstra's Attack on RSA with Chinese Remainder Algorithm

Boneh, DeMillo and Lipton [6] gave an attack on RSA implemented with the CRA. Their attack requires two signatures of a given message: one correct signature and one faulty signature. Lenstra independently worked out a similar attack against RSA with CRA which requires only one faulty signature of a known message [9]. In the following, we briefly outline Lenstra's attack.

The signature s of a message m equals $m^d \bmod n$ and thus $s^e \bmod n$ is again equal to $m \bmod n$. It is well known that s can be computed by computing

$$u \equiv m^d \bmod p \text{ and } v \equiv m^d \bmod q,$$

and by combining u and v using the CRA. If a fault occurs in the course of the computation of the signature, the resulting value, denoted as s', will most likely not satisfy $m \equiv s'^e \bmod n$. If, however, the fault occurred only during the computation of say, u, and if v and the CRA were carried out correctly, then the resulting faulty signature s' satisfies $s'^e \equiv m \bmod q$, but the same congruence mod p does not hold. Therefore, q divides $s'^e - m$ but p does not divide $s'^e - m$, so that a factor of n may be discovered by the recipient of the faulty signature s' by computing the greatest common divisor of n and $s'^e - m$. This attack is very powerful since it requires only one faulty signature and it works under a general fault model.

2.4 Some Possible Countermeasures

There may be a variety of attacks to PKCs by inducing faults. The means of breaking a PKC can be devised to be dependent on the specific PKC algorithm as well as on its implementation. Generally speaking, countermeasures to such attacks are relatively insensitive to both the implementation of a PKC and the attacking scenarios. Here we envisage two general approaches to counter such attacks, one is based on the principle of "check and balance" and the other based on the principle of "information hiding". The former can be done by checking/verifying the result before sending it to the outside world and the latter can be achieved by introducing some randomness in the intermediate stages of the cryptographic computation.

a) The attacks may be avoided by calculating the output 2 times and matching the two results. However, this approach doubles the computational time. As pointed out in [4], this double computation method also avoids their attack. The weakness of this counter measure is that it slows down the computation by a factor of 2, which is "not accepted for some applications"[4].

b) In many cases, the encryption key e is usually small. So we can verify the result by checking $m'^e = c \bmod n$? It is much more efficient than the double computation approach if e is small. This approach was also pointed out independently by Lenstra.

c) In some protocols for digital signature, a random string is chosen by the smart card and concatenated to a message m which is to be signed by the smart card. For example, m is a 412 binary string given to the smart card. The smart card randomly chooses a 100 bit number r and the output is $(m|r)^d \bmod n$. Since r is different each time, the attack does not work in such case.

d) In the case where e is large and where the tamper resistant device is required to compute $c^d \bmod n$, the following efficient method may be used to counter the attack. The tamper resistant device generates a random number r and computes $r^d \bmod n$. This can be done in advance, i.e., before c is input and when the device is idle. To compute $c^d \bmod n$, the device first computes $rc \bmod n$, then $(rc)^d \bmod n$, and finally $\frac{(rc)^d}{r^d} \bmod n$. If no fault takes place, the output is obviously correct. If any fault takes place, the output is masked by r. Since r is unknown to the attacker and different for every decryption, our attack does not work. For the example, in the case of Attack II, if d_i is 0 and d_i' is 1, then $m'/m \equiv r^{2^i} c_i \bmod n$. Since r is unknown to the attacker, the ratio is useless to him.

It should be pointed out that a) - d) work against our attacks while only a) c) work against Lenstra's attack.

3 Attacking Discrete Logarithm Based Schemes

The general concept of attacking the RSA scheme can be applied to attack against discrete logarithm based public key cryptosystems. In the following, we show our attacks to the ElGamal signature scheme, the Schnorr signature scheme, and the DSA. Throughout this section, we will denote a signer's private key as x and its binary representation as $x_{t-1}|x_{t-2}\cdots|x_i|\cdots x_1|x_0$, where t is the number of bits in x and x_i is the ith bit of x. The private key is kept inside a tamper resistant device and the corresponding public key can be made available to everyone.

The general steps followed by an attacker are as follows: 1) the attacker applies external physical effects to induce some bit errors at random locations in x and then obtains a faulty signature, 2) he performs some computations on the faulty signature to uncover part of the private key x. The attacker repeats steps 1) and 2) until he uncovers the binary representation of x or a sufficiently large number of bits that allow him to discover the rest of x by brute force. To keep the paper compact, we will only show steps 1) and 2) in the following, without explicitly showing the loops of the attack.

To simplify the description, we first show the attacks for the case of single bit error. We then briefly discuss the case of multiple bit errors.

3.1 Attacking the ElGamal Signature Scheme

In the ElGamal signature scheme [7], to generate a private and public key pair, we first choose a prime p, and two random numbers, g and x, such that both g and x are less than p. The private key is x and the public key is $(y \equiv g^x \bmod p, g, p)$.

To generate a signature on a message m, the signer first picks a random k such that k is relatively prime to $p - 1$. She then computes

$$w \equiv g^k \bmod p \text{ and } s \equiv (m - xw)/k \bmod (p - 1)$$

The signature is the pair w and s. To verify the signature, the verifier confirms that

$$y^w w^s \equiv g^m \bmod p.$$

Assume that x_i in x is changed to its complement x_i' during the process of signing of a message m. We denote the corrupted x as x' due to the flip of x_i. Then the outputs of the device will be

$$w \equiv g^k \bmod p \text{ and } s' \equiv (m - x'w)/k \bmod (p - 1)$$

Using w, s', m, and the signer's public key (y, p, g), the attacker computes

$$T \equiv y^w w^{s'} \bmod p \equiv g^m g^{w(x-x')} \bmod p.$$

Let $R_i \equiv g^{w2^i} \bmod p$ for $i = 0, 1, 2, ..., t - 1$. Then, we have

$$TR_i \equiv g^m \bmod p, \text{ if } x_i = 0$$

(since for $x_i = 0$ we have $x - x' = -2^i$) and

$$\frac{T}{R_i} \equiv g^m \bmod p, \text{ if } x_i = 1$$

(since for $x_i = 1$ we have $x - x' = 2^i$). The attacker computes TR_i and T/R_i and tests to see if either TR_i or T/R_i equals $g^m \bmod p$, for $i = 0, 1, ..., t - 1$. If a match is found, then one bit of x is found.

3.2 Attacking the Schnorr Signature Scheme

In the Schnorr signature scheme [11], to generate a private and public key pair, we first choose two primes, p and q, such that $p = zq + 1$ for a reasonably large q. We then select a number g not equal to 1, such that $g^q \equiv 1 \bmod p$. The signer's private key is a random x less than q, and the public key is $(y \equiv g^{-x} \bmod p, g, p, q)$.

To generate a signature on a message m, the signer first picks a random k that is less than q. She then computes

$$w \equiv g^k \bmod p, e = h(m|w) \text{ and } s \equiv ex + k \bmod q,$$

where h is a secure one-way hash function that outputs a number less than q. The signature is the pair e and s. Because

$$(g^s y^e \bmod p) = w,$$

to verify the signature, the verifier confirms that

$$h(m|(g^s y^e \bmod p)) = e$$

During the computation of s, assuming that x_i in x is flipped to x_i' and denote the corrupted x as x'. Then the outputs of the device will be

$$e \equiv h(m|w) \bmod p \text{ and } s' \equiv ex' + k \bmod q$$

Using e, s', m, and the signer's public key (y, p, g, q), the attacker computes

$$T \equiv g^{s'} y^e \equiv wg^{e(x'-x)} \bmod p$$

Let $R_i \equiv g^{e2^i} \bmod p$ for $i = 0, 1, 2, ..., t-1$. It is easy to see that $TR_i \equiv wg^{e(x'-x+2^i)} \bmod p$ and $T/R_i = wg^{e(x'-x-2^i)} \bmod p$. Then we have

$$h(m|(TR_i \bmod p)) = e, \text{ if } x_1 = 1$$

(since for $x_i = 1$, we have $x' - x = -2^i$ and then $TR_i \equiv w \bmod p$), and

$$h(m|(T/R_i \bmod p)) = e, \text{ if } x_i = 0$$

(since for $x_i = 0$, we have $x' - x = 2^i$ and then $T/R_i \equiv w \bmod p$). Therefore, by iterating through different i and matching e with $h(m|(TR_i \bmod p))$ and $h(m|(T/R_i \bmod p))$, the attacker can discover the ith bit, x_i, of the private key x.

In [6], Boneh, DeMillo and Lipton gave an attack against the Schnorr identification scheme [11]. In their attack, it was required that 1) the verifier uses the same challenge e (which plays the same role as the e in the Schnorr signature scheme) in all invocations of the identification protocol without being detected by the prover's tamper resistant device, and 2) a bit error is introduced in the random number k (which plays the same role as the k in the Schnorr signature scheme). Because of these two requirements, their attack can not be applied to break the Schnorr signature scheme. On the other hand, our attack to the Schnorr signature scheme can be applied to break the Schnorr identification scheme with little modification.

3.3 Attacking the DSA

In the DSA, to generate a private and public key pair, we first choose a prime p such that $p = zq + 1$ for a reasonably large prime q. We then compute $g \equiv b^{(p-1)/q} \bmod p$, where b is any number less than $p-1$ such that $(b^{(p-1)/q} \bmod p)$ is greater than 1. The signer's private key is x, a random number less than q, and the public key is $(y \equiv g^x \bmod p, g, p, q)$.

To sign a message m, the signer first picks a random k that it is less than q. She then computes

$$w \equiv g^k \bmod p \bmod q \text{ and } s \equiv (e + wx)/k \bmod q,$$

where $e = h(m)$ with h being a secure one-way hash function that outputs a number less than q. The signature is the pair w and s. To verify the signature, the verifier confirms that

$$w \equiv g^{(ue \bmod q)} y^{(uw \bmod q)} \bmod p \bmod q,$$

where $u \equiv 1/s \bmod q$.

The attacker applies external physical effects to the tamper resistant device and at the same time asks the device to sign a message m. During the process of calculating s, we assume that the ith bit of x is changed from x_i to its complement x_i'. Let x' denote the corrupted x due to the flip of x_i. Then the outputs of the device will be

$$w \equiv g^k \bmod p \bmod q \text{ and } s' \equiv (e + wx')/k \bmod q$$

Using w, $u' \equiv 1/s' \bmod q$, m, and the signer's public key (y, p, g, q), the attacker can compute $e = h(m)$ and

$$T \equiv g^{(u'e \bmod q)} y^{(u'w \bmod q)} \equiv g^{(u'(e+xw)) \bmod q} \bmod p \bmod q.$$

Let $R_i \equiv g^{(u'w2^i \bmod q)} \bmod p \bmod q$ for $i = 0, 1, 2, ..., t - 1$. Then we have
$$TR_i \equiv g^{(u'(e+w(x+2^i))) \bmod q} \bmod p \bmod q$$
$$T/R_i \equiv g^{(u'(e+w(x-2^i))) \bmod q} \bmod p \bmod q.$$
It is easy to show that

$$TR_i \equiv w \bmod p \bmod q, \text{ if } x_i = 0 \qquad T/R_i \equiv w \bmod p \bmod q, \text{ if } x_i = 1$$

So by iterating through different i and matching w with $TR_i \bmod p \bmod q$ and $T/R_i \bmod p \bmod q$, the attacker can discover the value of x_i.

3.4 Multiple Bit Errors

Extension of the above methods in attacking discrete log based digital signature schemes to the cases of multiple faults in x is straightforward. In the case of single fault, the Rs, denoted as R_i in the above single bit error case, each has a single argument i. Their computation and subsequent comparison is of complexity $O(2t)$. In the case of $j > 1$ faults in x, the Rs, which we will denote as $R_{i1,i2,...,ij}$, will each have j arguments and their computations and subsequent comparisons are of complexity $O((2t)^j)$.

3.5 Countermeasures

The countermeasure achieved through double computation as mentioned in section 2.4 also applies to the attacks presented in this section.

Another countermeasure to the attack against discrete log based signature schemes is that the tamper resistant device stores both x and $1/x$, where x is used in the computation of s and $1/x$ is used to check the correctness of s. As an example let's consider the Schnorr signature scheme. Right after computing $s' \equiv ex' + k \bmod q$, the device verifies the value of s' by comparing e with $(s' - k)(1/x) \bmod q$. If these two values are the same, the result is considered correct; otherwise, the device is reset.

In general, to prevent corrupted variables from being used in a calculation and subsequently causing breaking of a cryptosystem, we suggest that the variable and its inverse be stored somewhere before the calculation takes place. The variable is used for the calculation and its inverse can be used to verify the result of the calculation.

To illustrate the above concept, let's again consider the Schnorr signature scheme. Suppose we want to make sure that the correct values of both x and k are used in the calculation of $s = ex + k$. The tamper resistant device stores $x, 1/x, k, 1/k$ somewhere before the calculation starts. After computing $s' = ex' + k'$, the device checks to see if $e = k(s'(1/k) - 1)(1/x)$. The value s' is considered correct only if the equality holds.

4 Concluding Remarks

The attack to public-key cryptographic schemes on tamper resistant devices presented in this paper makes use of transient faults. Our attacking model is independent of the implementation of a specific cryptosystem and seems to be applicable to breaking large classes of public-key cryptosystems. In particular, we showed how to break the RSA, the ElGamal signature scheme, the Schnorr signature scheme, and the DSA.

This attack highlighted that hardware faults can cause fatal leakage of the private/secret key values and may eventually lead to breaking of a cryptosystem. Therefore, it is important to take fault tolerance into serious consideration in the design of cryptosystems and to strike a balance between low overhead and high robustness. As a first step, we have proposed some methods to counter our attack. It should be noted that there are many other ways of breaking tamper resistant devices in addition to the ones outlined in this paper. In general, design of fault tolerant tamper resistant devices is a very challenging problem. Readers interested in getting more information in this area are refereed to the excellent paper by Anderson and Kuhn [1].

5 Acknowledgments

The work reported in this paper was motivated by the Bellcore Press Release [4] and further motivated by Biham and Shamir's research announcement [5].

The fundamental concept of breaking cryptosystems in the presence of hardware faults released in [4] was the main driving force behind our research. The authors would also like to thank Dr. Arjen Lenstra for his valuable comments on an earlier version of our manuscript.

References

1. R. Anderson and M. Kuhn, "Tamper Resistance - A Cautionary Note", to appear in the Proceedings of the 2nd Workshop on Electronic Commerce, Oakland, CA., Nov. 18-20, 1996.
2. F. Bao, R. Deng, Y. Han, A. Jeng, D. Narasimhalu, and T. Ngair, "Another New Attack to RSA on Tamperproof Devices", 23rd October. 1996, http://www.itd.nrl.navy.mil/ITD/5540/ieee/cipher/news-items/961022.sgtamper.html; "A Method to Counter Another New Attack to RSA on Tamperproof Devices", 24th October. 1996, http://www.itd.nrl.navy.mil/ITD/5540/ieee/cipher/news-items/ 961024.sgtampercounter.html.
3. F. Bao, R. Deng, Y. Han, A. Jeng, D. Narasimhalu, and T. Ngair, "New Attacks to Public Key Cryptosystems on Tamperproof Devices", 29th October. 1996,http://www.itd.nrl.navy.mil/ITD/5540/ieee/cipher/news-items/.
4. Bellcore Press Release, "New Threat Model Breaks Crypto Codes", Sept. 1996, http://www.bellcore.com/PRESS/ADVSRY96/facts.html.
5. E. Biham and A. Shamir,"Research Announcement: A New Cryptanalytic Attack on DES", 18th October 1996, http://jya.com/dfa.htm.
6. D. Boneh, R. A. DeMillo, and R. J. Lipton, "On the Importance of Checking Computations", Submitted to Eurocrypt 96.
7. T. ElGamal, "A Public-Key Cryptosystems and a Signature Scheme Based on Discrete Logarithms", IEEE Trans. Information Theory, Vol. IT-31, No. 4, 1985, pp. 469-472.
8. M. Joye and J.-J. Quisquater, "Attacks on systems using Chinese remaindering", Technical Report CG-1996/9 of UCL, http://www.dice.ucl.ac.be/crypto/.
9. A. K. Lenstra, "Memo on RSA Signature Generation in the Presence of Faults", Manuscript, Sept. 28, 1996. Available from Author at arjen.lenstra@citicorp.com.
10. R. L. Rivest, A. Shamir, and L. M. Adleman,"A Method for Obtaining Digital Signatures and Public-Key Cryptosystems", Communications of the ACM, vol. 21, No. 2, Feb. 1978, pp. 120-126.
11. C. Schnorr, "Efficient Signature Generation by Smart Cards", J. Cryptology, Vol. 4, 1991, pp. 161-174.

Low Cost Attacks on Tamper Resistant Devices

Ross Anderson[1], Markus Kuhn[2]

[1] Computer Laboratory, Pembroke Street, Cambridge CB2 3QG, UK
rja14@cl.cam.ac.uk
[2] COAST Laboratory, Purdue University, West Lafayette, IN 47907, USA
kuhn@cs.purdue.edu

Abstract. There has been considerable recent interest in the level of tamper resistance that can be provided by low cost devices such as smartcards. It is known that such devices can be reverse engineered using chip testing equipment, but a state of the art semiconductor laboratory costs millions of dollars. In this paper, we describe a number of attacks that can be mounted by opponents with much shallower pockets.

Three of them involve special (but low cost) equipment: differential fault analysis, chip rewriting, and memory remanence. There are also attacks based on good old fashioned protocol failure which may not require any special equipment at all. We describe and give examples of each of these. Some of our attacks are significant improvements on the state of the art; others are useful cautionary tales. Together, they show that building tamper resistant devices, and using them effectively, is much harder than it looks.

1 Introduction

An increasing number of large and important systems, from pay-TV through GSM mobile phones and prepayment gas meters to smartcard electronic wallets, rely to a greater or lesser extent on the tamper resistance properties of smartcards and other specialist security processors.

This tamper resistance is not absolute: an opponent with access to semiconductor test equipment can retrieve key material from a smartcard chip by direct observation and manipulation of the chip's components. It is generally believed that, given sufficient investment, any chip-sized tamper resistant device can be penetrated in this way. A number of less expensive techniques for attacking specific tamper resistant devices are also known [2].

So the level of tamper resistance offered by any particular product can be measured by the time and cost penalty that the protective mechanisms impose on the attacker. Estimating these penalties is clearly an important problem, but is one to which security researchers, evaluators and engineers have paid less attention than perhaps it deserves. (The relatively short bibliography at the end of this article bears witness to that.)

We will adopt the taxonomy of attackers proposed by IBM to guide designers of security systems that rely to some extent on tamper resistance [1]:

Class I (clever outsiders): They are often very intelligent but may have insufficient knowledge of the system. They may have access to only moderately sophisticated equipment. They often try to take advantage of an existing weakness in the system, rather than try to create one.

Class II (knowledgeable insiders): They have substantial specialized technical education and experience. They have varying degrees of understanding of parts of the system but potential access to most of it. They often have highly sophisticated tools and instruments for analysis.

Class III (funded organisations): They are able to assemble teams of specialists with related and complementary skills backed by great funding resources. They are capable of in-depth analysis of the system, designing sophisticated attacks, and using the most advanced analysis tools. They may use Class II adversaries as part of the attack team.

In this paper, we present and develop a number of techniques that can make smartcards and other tamper resistant devices vulnerable to class II or even class I attackers.

2 Differential Fault Analysis

In [5], Biham and Shamir announced an attack on DES based on 200 ciphertexts in which one-bit errors have been induced by environmental stress. The fault model they used had been proposed by Boneh and others in [11] and its effects investigated further in [16, 10]. It assumes that by exposing a processor to a low level of ionising radiation, or some other comparable insult, that one-bit errors can be induced in the data used and specifically in the key material fed into the successive rounds.

In [6], it is shown how this method could be extended to reverse engineer algorithms whose structure is unknown. In each case, the critical observation is that errors that occur in the last few rounds of the cipher leak information about the key, or algorithm structure, respectively. In [7], a number of further results are given; if faults can be induced in the last one or two rounds of the algorithm, then fewer faulty ciphertexts are needed. In [11], it was shown that on a similar fault model, attacks could be carried out on public key systems. In particular, an RSA modulus could be factored given a number of faulty signatures.

The problem with these proposed attacks is that no-one has demonstrated the feasibility of the fault model. Indeed, with many security processors, the key material is held in EEPROM together with several kilobytes of executable code; so it is likely that a random one-bit error which did have an effect on the device's behaviour would be more likely to crash the processor or yield an uninformative error than to produce a faulty ciphertext of the kind required for the above attacks.

In this section, we show that a different, and more realistic, fault model gives significantly better attacks. In the following sections, we will discuss some other faults that can be induced by low budget attackers and show that they, too, lead to feasible attacks. Many of these attacks can also be extended to cases in which we do not initially know the algorithm in use, or where our knowledge of the system is otherwise imperfect. They fall squarely within the definition of what a class I opponent might do.

2.1 A Realistic Differential Attack

In [2], we discussed an attack that has been used by amateur hackers in assaults on pay-TV smartcards. The idea is to apply a glitch — a rapid transient — in either the clock or the power supply to the chip. Typical attacks had involved replacing a 5 MHz clock pulse to a smartcard with one or more 20 MHz pulses. Because of the different number of gate delays in various signal paths and the varying RC parameters of the circuits on the chip, this affects only some signals, and by varying the precise timing and duration of the glitch, the CPU can be made to execute a number of completely different wrong instructions. These will vary from one instance of the chip to another, but can be found by a systematic search using hardware that can be built at home.

We do not claim to have invented this attack; it appears to have originated in the pay-TV hacking community, which has known about it for at least a year. In the form described in [2], it involved a loop that writes the contents of a limited memory range to the serial port:

```
1  b = answer_address
2  a = answer_length
3  if (a == 0) goto 8
4  transmit(*b)
5  b = b + 1
6  a = a - 1
7  goto 3
8  ...
```

The idea is to find a glitch that increases the program counter as usual but transforms either the conditional jump in line 3 or the loop variable decrement in line 6 into something else. Then, by repeating the glitch, the entire contents of memory can be dumped.

When applied at the algorithm level rather than at the level of control code, this attack is also highly effective, as we shall now see. The import of this work is that attacks based on inducing errors in instruction code are easier, and more informative, than attacks based on inducing errors in data.

2.2 Attacking RSA

The Lenstra variant of the attack on RSA goes as follows: if a smartcard computes an RSA signature S on a message M modulo $n = pq$ by computing it modulo p and q separately and then combining them using the Chinese Remainder Theorem, and if an error can be induced in either of the former computations, then we can factor n at once. If e is the public exponent, and the 'signature' $S = M^d(\bmod\ pq)$ is correct modulo p but incorrect modulo q, then we will have

$$p = \gcd(n, S^e - M) \tag{1}$$

This is ideal for a glitch attack. As the card spends most of its time calculating the signature mod p and mod q, and almost any glitch that affects the output will do, we do not have to be selective about where in the instruction sequence the glitch is applied. Since only a single signature is needed, the attack can be performed online: a Mafia shop's point-of-sale terminal can apply the glitch, factor the modulus, calculate what the correct signature should be, and send this on to the bank, all in real time.

Thus the Mafia can harvest RSA secret keys without the customer or his bank noticing anything untoward about the transaction performed at their shop. Given that implementers of the new EMV electronic purse system propose to have only 10,000 different RSA secret keys per issuing bank [14], the Mafia may soon be able to forge cards for a substantial proportion of the user population.

2.3 Attacking DES

When we can cause an instruction of our choice to fail, then there are several fairly straightforward ways to attack DES. We can remove one of the 8-bit xor operations that are used to combine the round keys with the inputs to the S-boxes from the last two rounds of the cipher, and repeat this for each of these key bytes in turn. The erroneous ciphertext outputs that we receive as a result of this attack will each differ from the genuine ciphertext in the output of usually two, and sometimes three, S-boxes. Using the techniques of differential cryptanalysis, we obtain about five bits of information about the eight keybits that were not xor'ed as a result of the induced fault. So, for example, six ciphertexts with faulty last rounds should give us about 30 bits of the key, leaving an easy keysearch.

An even faster attack is to reduce the number of rounds in DES to one or two by corrupting the appropriate loop variable or conditional jump, as in the protocol attack described above. Then the key can be found by inspection. The practicality of this attack will depend on the implementation detail.

Thus DES can be attacked with somewhere between one and ten faulty ciphertexts. But how realistic is it to assume that we will be able to target particular instructions?

In most smartcards, the manufacturer supplies a number of routines in ROM. Though sometimes presented as an 'operating system', the ROM code is more of a library or toolkit that enables application developers to manage communications and other facilities. Its routines usually include the DES algorithm (or a proprietary algorithm such as Telepass), and by buying the manufacturer's smartcard development toolkit (for typically a few thousand dollars) an attacker can get full documentation plus real specimens for testing. In this case, individual DES instructions can be targeted.

When confronted with an unfamiliar implementation, we may have to experiment somewhat (we have to do this anyway with each card in order to find the correct glitch parameters). However the search space is relatively small, and on looking at a few DES implementations it becomes clear that we can usually recognise the effects of removing a single instruction from either of the last two rounds. (In fact, many of these instructions yield almost as much information when removed from the implementation as the key xor instructions do.)

2.4   Reverse Engineering an Unknown Block Cipher

We can always apply clock and power glitches until simple statistical tests suddenly show a high dependency between the input and output bits of the encryption function, indicating that we have succeeded in reducing the number of rounds. This may be practical even where the implementation details are unknown, which leads us to ask whether we can use our attack techniques to reverse engineer an unknow algorithm, such as Skipjack, without needing to use expensive chip testing equipment.

In [6, 7], Biham and Shamir discuss this problem in their fault model of one-bit random data errors. As before, they identify faults that affected only the last round or rounds; this can be done by looking for ciphertexts at a low Hamming distance from each other. They then identify which output bits correspond to the left and right halves, and next look at which bits in the left half are affected by one bit changes in the last-but-one right half. In the case of a cipher such as DES with S-boxes, the structure will quickly become clear and with enough ciphertexts the values of the S-boxes can be reconstructed. They report that with 500 ciphertexts the gross structure can be recovered, and with about 10,000 the S-box entries themselves can be found.

Our technique of causing faults in instructions rather than in data bits is more effective here, too. We can attack the last instruction, then the second last instruction, and so on. The number of ciphertexts required for this attack is about the same as for Biham and Shamir's.

Let us now consider an actual classified algorithm. 'Red Pike' was designed by GCHQ for encrypting UK government traffic classified up to 'Restricted', and the Department of Health wishes to use it to encrypt medical records. The British Medical Association preferred that an algorithm be chosen that had been in the open literature for at least two years and had withstood serious attempts to find shortcut attacks (3DES, Blowfish, SAFER K-128, WAKE, ...).

In order to try and persuade the BMA that Red Pike was sound, the government commissioned a study of it by four academics [18]. This study states that Red Pike 'uses the same basic operations as RC5' (p 4); its principal operations are add, exclusive or, and left shift. It 'has no look-up tables, virtually no key schedule and requires only five lines of code' (p 4). Other hints include that 'the influence of each key bit quickly cascades' (p 10) and 'each encryption involves of the order of 100 operations' (p 19).

We can thus estimate the effort of reverse engineering Red Pike from a tamper resistant hardware implementation by considering the effort needed to mount a similar attack on RC5 [19].

Removing the last operation — the addition of key material — yields an output in which the right hand side is different (it is (B xor A) shl A where A and B are the left and right halves respectively). This suggests, correctly, that the algorithm is a balanced Feistel cipher without a final permutation. Removing the next operation — the shift — makes clear that it was a 32 bit circular shift but without revealing how it was parametrised. Removing the next operation — the xor — is transparent, and the next — the addition of key material in the previous round — yields an output with the values A and B in the above expression. It thus makes the full structure of the data-dependent rotation clear. The attacker can now guess that the algorithm is defined by

```
A = ((A xor B) shl B) op key
B = ((B xor A) shl A) op key
```

Reverse engineering RC5's rather complex key schedule (and deducing that 'op' is actually +) would require single-stepping back through it separately; but if we guess that 'op' is +, we can find the round key bits directly by working back through the rounds of encryption.

So, apart from its key schedule, RC5 may be about the worst possible algorithm choice for secret-algorithm hardware applications, where some implementations may be vulnerable to glitch attacks. If Red Pike is similar but with a simpler key schedule, then it could be more vulnerable still. However, since the government plans to make Red Pike available eventually in software, this is not a direct criticism of the design or choice of that algorithm.

It does all suggest, though, that secret-hardware algorithms should be more complex; large S-boxes kept in EEPROM (that is separate from the program memory) may be one way of pushing up the cost of an attack. Other protective measures that prudent designers would consider include error detection, multiple encryption with voting, and designing the key schedule so that the keys from a small number of rounds are not enough for a break.

3   Chip Rewriting Attacks

Where the implementation is familiar, there are a number of ways to extract keys from the card by targeting specific gates or fuses or by overwriting specific memory locations. Bovenlander has described breaking smartcards by using two microprobe needles to bridge the fuse blown at the end of the card test cycle, and using the re-enabled test routine to read out the memory contents [12]. Even where this is not possible, memory cells can be attacked; this can also be done on a relatively modest budget.

3.1 ROM overwrite attacks

Single bits in a ROM can be overwritten using a laser cutter microscope, and where the DES implementation is well known, we can find one bit (or a small number of bits) with the property that changing it will enable the key to be extracted easily. The details will depend on the implementation but we might well be able, for example, to make a jump instruction unconditional and thus reduce the number of rounds to one or two. We can also progressively remove instructions such as exclusive-or's of key material.

Where we have incomplete information on the implementation, ROM over-writing attacks can be used in other ways. For example, DES S-boxes in ROM can be identified and a number of their bits overwritten such that the encryption function becomes a linear transformation over GF(2); we can then extract the key from a single plaintext/ciphertext pair.

3.2 EEPROM modification attacks

Where the algorithm is kept in EEPROM, we can use two microprobing needles to set or reset the target bit [17]. We can use this technique to carry out the above attacks; but the fact that we can both set and reset bits opens up still more opportunities.

Recall that the DES algorithm uses keys with odd parity, and a proper implementation will require that a key with the wrong parity will cause an error message to be returned (the VISA security module described below is an example of such equipment). Suppose further that we know the location of the DES key in memory but cannot read it directly; this could well be the case where the key is kept in EEPROM at a known location (smartcard software writers often locate keys at the bottom end of EEPROM memory), but we lack the equipment to carry out the attacks described in [2]. We can proceed as follows.

Set the first bit of the EEPROM containing the target DES key to 1 (or 0, the choice doesn't matter) and operate the device. If it still works, the keybit was a 1. If you get a 'key parity error' message, then the bit was zero. Move on to the next bit; set it to 1 and see if this changes the device's response (from encryption to error or vice versa). Even where the protocol uses some form of key redundancy that we do not understand, we can react to error messages by simple changing the keybit back to its original value.

Both microprobes and laser cutter microscopes are often found in universities — the former in electrical engineering departments, and the latter in cellular biology laboratories. Undergraduates can obtain unsupervised access to them; other class I attackers can purchase them for at most a few thousand dollars.

3.3 Gate destruction attacks

At the rump session of the 1997 workshop on Fast Software Encryption, Eli Biham and Adi Shamir presented a novel and interesting attack on DES. The idea is to use a laser cutter to destroy an individual gate in the hardware implementation of a known block cipher.

The example they gave was DES, which is typically implemented with hardware for a single round, plus a register that holds the output of round k and sends it back as the input to round $k+1$. Biham and Shamir pointed out that if the least significant bit of this register is stuck, then the effect is that the least significant bit of the output of the round function is set to zero. By comparing the least significant six bits of the left half and the right half, several bits of key can be recovered; given about ten ciphertexts from a chip that has been damaged in this way, information about the round keys of previous rounds can be deduced using the techniques of differential cryptanalysis, and enough of the key can be recovered to make keysearch easy.

This is an extremely impressive attack, and in fact the first one that works against ciphers such as DES when the plaintext is completely unknown. (This is the case in many smartcard applications where the card uses successive payment transactions to report its internal state to the issuer.)

We observe that there is a simple countermeasure to this new attack: a chip modified in this way will have the property that encryption and decryption are no longer inverses. So a simple self-test procedure can be added that takes an arbitrary input, encrypts and decrypts under an arbitrary key, and compares the result with the original block. (This test is already being implemented by one of our clients in a chip currently under development.)

4 Memory Remanence Attacks

In a brilliant USENIX paper [15], Gutman described the mechanisms that cause both static and dynamic RAM to 'remember' values that they have stored for a long period of time. A prudent security engineer will ask what the effect of this is in the real world.

We looked at a security module used in a bank. Many banks use a system devised by IBM and refined by VISA to manage the personal identification numbers (PINs) issued to customers for use with automatic teller machines [4]. The PIN is derived from the account number by encrypting it with a 'PIN key', decimalising the result and adding a decimal 'offset' (without carry) to get the PIN the customer must enter. (The offset's function is to enable the customer to choose his own PIN.) An example of the calculation is [4]:

```
Account number:        8807012345691715
PIN key:               FEFEFEFEFEFEFEFE
Result of DES:         A2CE126C69AEC82D
Result decimalised:    0224126269042823
Natural PIN:           0224
Offset:                6565
Customer PIN:          6789
```

The function of the security module is to perform all these cryptographic operations, plus associated key management routines, in trusted hardware, so as to support a dual control policy: no single member of any bank's staff should have access to a customer PIN [20]. Thus, for example, the module will only perform a 'verify PIN' command if the PIN is supplied encrypted under a key allocated to an automatic teller machine or to a corresponding bank. In this way, bank programmers are prevented from using the security module as an oracle to perform exhaustive PIN search.

In order to enforce this, the security module needs to be able to mark keys as belonging to a particular functionality class. It does this by encrypting them with 3DES under one of 12 pairs of DES master keys that are stored in low memory. Thus for example ATM keys are stored encrypted under master keys 14 and 15, while the working keys used to communicate with other banks are stored encrypted under master keys 6 and 7. The encrypted values of long term keys such as the PIN key are often included inline in application code and are thus well known to the bank's programming staff.

So security depends on the module's tamper resistance, and this is provided for by lid switches that cut power to the key memory when the unit is opened (it needs servicing every few years to change the battery). Master keys are loaded back afterwards in multiple components by trusted bank staff.

We looked at one such device, which dated from the late 1980's, and found that the master key values were almost intact on power-up. The number of bits in error was of the order of 5-10%. We cannot give more accurate figures as we were not permitted to copy either the correct master key values, nor the almost-correct values that had been 'burned in' to the static RAM chips. We are also not permitted to name the bank at which these modules are installed, and do not consider it prudent to name their manufacturer.

This level of memory remanence would be alarming enough. However, it has a particularly pernicious and noteworthy interaction with DES key parity in this common application.

If each DES key is wrong by five bits, then the effort involved in searching for the 10 wrong bits in a double DES key might be thought to be 112-choose-10 operations. Each operation would involve (a) doing a 2-key 3DES decryption of a 56 bit PIN key whose enciphered value is, as we noted, widely known (b) in the 2^{-8} of cases where this result has odd parity, enciphering an account number with this as a DES key to see if the (decimalised) result is the right PIN. The effort is about 3 times 112-choose-10 DES operations — say 2^{50}. But it would probably be cheaper to do a hardware keysearch on the PIN key directly than to try to implement this more complex 2^{50} search in either hardware or software.

However, the bytewise nature of the DES key redundancy reduces the effort by several orders of magnitude. If no key byte has a double error, then the effort is seven tries for each even parity byte observed, or 3 times 7^{10} — about 2^{30}, which is easy. If there is one key byte with a double error, the effort is 2^{38}, giving a search of 2^{40} DES operations — which is feasible for a class I attacker.

This is not the first instance of DES parity being a hindrance rather than a help. In one case, the Kerberos partity-checking DES implementation was grafted into an encrypting telnet implementation that derived its key material from a Diffie-Hellman exchange. As the Diffie-Hellman key bits were random, only 1 in 256 exchanges resulted in a legal key; in all other cases, key loading failed and (as the implementation didn't check the return code) the session would continue using an uninitialised key [8]. In another, a misunderstanding led to the PIN key used by a number of banks being chosen as a password made up of ASCII characters with odd parity; and in other applications, it is common for an ASCII password to have parity set and be used as a key. The problem here is that ascii characters have a zero in the high order bit, while DES parity operates on the low order bit; so the key diversity is less than 2^{48}, and in fact is even less than the entropy of the chosen passwords [13].

5 Protocol Failure

Poorly designed protocols are a more common source of attacks than many people recognise [3]. Many of them also require only very simple and cheap equipment to exploit.

For example, satellite TV decoders typically have a hardware cryptoprocessor that deciphers the video signal, and a microcontroller which passes messages between the cryptoprocessor and the customer smartcard that contains the key material. If a customer stops paying his subscription, the system typically sends a message over the air which instructs the decoder to disable the card. In the 'Kentucky Fried Chip' hack, the microcontroller was replaced with one which blocked this particular message [3].

Another example is given in [2], which describes an attack on the Dallas Semiconductor DS5002FP secure microcontroller. This attack utilises a protocol failure to circumvent the encryption system used to protect off-chip memory.

Some protocol failures require no equipment at all to exploit, and one example that has come to our attention arose from a modification made to a bank security module.

One bank was upgrading its systems and wished to change the format of its customer account numbers. Changing the numbers meant that the 'natural PIN' calculated from the account number would change. But the bank did not wish to inconvenience its customers by forcing new PINs on them; so it decided to calculate suitable offsets so that the customer PIN would be unchanged from one card generation to the next.

The security module as supplied did not support such a transaction (for reasons that will shortly become clear). The manufacturer was duly contacted and asked to provide it; modified software was duly supplied, but with a warning that this should only be used for a batch run to calculate the necessary offsets, and then discarded, as it was dangerous. But the nature of the danger was not spelled out, and due to personnel changes and project delays the card number change could not be carried out at once. The effect was that the modified software was installed and left in place.

About a year later, one of the bank's programmers noticed a simple attack. The additional transaction had the syntax: 'given an initial account number of X and offset of Y, calculate an offset which will enable this PIN to be used on account number Z'. The programmer could input the account number and offset of a target as X and Y (the majority of offsets were zero in any case) and his own account number as Z. The returned value enabled him to trivially calculate the target's PIN.

Fortunately for the bank, the programmer brought this vulnerability to the attention of authority, rather than exploiting it.

6   Conclusions

We have improved on Differential Fault Analysis. Rather than needing about 200 faulty ciphertexts to recover a DES key, we need between one and ten. We can factor RSA moduli with a single faulty ciphertext. We can also reverse engineer completely unknown algorithms; this appears to be faster than Biham and Shamir's approach in the case of DES, and is particularly easy with algorithms that have a compact software implementation such as RC5. Unlike some previous work, our attacks use a realistic fault model, which has actually been implemented and can be used against fielded systems. The critical idea is to cause errors in code rather than in data.

We have also shown how low cost and commonly available laboratory equipment, such as microprobes and laser cutter microscopes, can be used to implement chip-surface attacks to recover key material from supposedly secure processors. Key redundancy, such as the key parity of DES, can be used to facilitate such attacks. This work reinforces the lesson from [9] — that key redundancy requires more careful consideration than has usually been accorded it in the past.

An example of this is that the particular form of key redundancy used in DES can interact quite lethally with the memory remanence properties of SRAM chips commonly used in banking security modules. The effect is that master key material can fairly easily be recovered from a discarded security module. Such devices should as a matter of policy be destroyed carefully.

Finally, in addition to attacks involving non-obvious interactions of protocol features with hardware features, there are many cases in which physical protection can be circumvented by pure protocol attacks. Tamper resistant devices are not only much harder to build than many people realise; they are also much harder to program and to use.

Acknowledgements

Mike Roe pointed out that the glitch attack on RSA can be done in real time by a Mafia owned point-of-sale terminal; Stefan Lucks the way to linearise DES.

References

1. DG Abraham, GM Dolan, GP Double, JV Stevens, "Transaction Security System", in *IBM Systems Journal* v 30 no 2 (1991) pp 206–229

2. RJ Anderson, MG Kuhn, "Tamper Resistance — a Cautionary Note", in *The Second USENIX Workshop on Electronic Commerce Proceedings* (Nov 1996) pp 1–11

3. RJ Anderson, RM Needham, "Programming Satan's Computer", in *'Computer Science Today'*, Springer Lecture Notes in Computer Science v 1000 pp 426–441

4. RJ Anderson, "Why Cryptosystems Fail", in *Proceedings of the 1st ACM Conference on Computer and Communications Security* (November 1993) pp 215–227

5. E Biham, A Shamir, "A New Cryptanalytic Attack on DES", preprint, 18/10/96

6. E Biham, A Shamir, "Differential Fault Analysis: Identifying the Structure of Unknown Ciphers Sealed in Tamper-Proof Devices", preprint, 10/11/96

7. E Biham, A Shamir, "Differential Fault Analysis: A New Cryptanalytic Attack on Secret Key Cryptosystems", preprint, 21/11/96

8. M Blaze, *personal communication*

9. M Blaze, "Protocol Failure in the Escrowed Encryption Standard", in *Proceedings of the 2nd ACM Conference on Computer and Communications Security* (2–4 November 1994), ACM Press, pp 59–67

10. F Bao, RH Deng, Y Han, A Jeng, AD Nirasimhalu, T Ngair, "Breaking Public Key Cryptosystems in the Presence of Transient Faults", *this volume*

11. D Boneh, RA DeMillo, RJ Lipton, "On the Importance of Checking Computations", preprint, 11/96

12. E Bovenlander, invited talk on smartcard security, Eurocrypt 97

13. P Farrell, *personal communication*

14. L Guillou, *comment from the floor of Crypto 96*

15. P Gutman, "Secure Deletion of Data from Magnetic and Solid-State Memory", in *Sixth USENIX Security Symposium Proceedings* (July 1996) pp 77–89

16. M Joye, F Koeune, JJ Quisquater, "Further results on Chinese remaindering", Université Catholique de Louvain Technical Report CG-1997-1, available at <http://www.dice.ucl.ac.be/crypto/tech_reports/CG1997_1.ps.gz>

17. O Kocar, "Hardwaresicherheit von Mikrochips in Chipkarten", in *Datenschutz und Datensicherheit* v 20 no 7 (July 96) pp 421–424

18. C Mitchell, S Murphy, F Piper, P Wild, "Red Pike — An Assessment", Codes and Ciphers Ltd 2/10/96

19. RL Rivest, "The RC5 Encryption Algorithm", in *Proceedings of the Second International Workshop on Fast Software Encryption* (December 1994), Springer LNCS v 1008 pp 86–96

20. *'VISA Security Module Operations Manual'*, VISA, 1986

Entity Authentication and Authenticated Key Transport Protocols Employing Asymmetric Techniques

Simon Blake-Wilson[1]* and Alfred Menezes[2]

[1] Dept. of Mathematics, Royal Holloway, University of London, Egham,
Surrey, TW20 0EX, United Kingdom. Email: phah015@rhbnc.ac.uk
[2] Dept. of Discrete and Statistical Sciences, Auburn University,
Auburn, AL 36849-5307, U.S.A. Email: ajmeneze@math.uwaterloo.edu

Abstract. This paper investigates security proofs for protocols that employ asymmetric (public-key) techniques to solve two problems: entity authentication and authenticated key transport.

A formal model is provided, and a definition of the goals within this model is supplied. Two protocols are presented and proven secure within this framework, given the existence of certain cryptographic primitives. The practical implementation of these protocols is discussed. We emphasize the relevance of these theoretical results to the security of systems used in practice. In particular, our results imply the security of some protocols standardized by ISO [15, 16] and NIST [20] in the model proposed.

This work is heavily influenced by the work of Bellare and Rogaway [1, 5], who demonstrate proven secure protocols for these problems using symmetric cryptosystems. Our paper is an extension of their work to the public-key setting.

1 Introduction

The *key transport problem* is stated as follows: one entity wishes to select keying information and communicate it in secret to another entity over a distributed network. If each entity also desires an assurance of the other's identity, this is known as *(mutual) authenticated key transport*. This keying information can then be used by the two entities to provide security services, just like a traditional secret key agreed face-to-face.

Solutions to the (authenticated) key transport problem come in various flavors. This paper concentrates on one particular flavor: those solutions which require the two entities to share only some authentic (but not secret) information in advance.

Since their introduction in the 1970's [10, 18], public-key techniques have been extensively used to solve the authenticated key transport problem. However none

---

* The author is an EPSRC CASE student sponsored by Racal Airtech. Work performed while a visiting student at Auburn University funded by the Fulbright Commission.

of these solutions has been provably demonstrated to achieve this goal, and this deficiency has led in many cases to the use of flawed protocols (see [19, 17]). The flaws have, on occasion, taken years to discover; at best, such protocols must be employed with the fear that a flaw will later be uncovered.

A closely related problem is the *entity authentication problem*. Here each communicating party merely desires an assurance of the other's identity.

The above problems are fundamental to the success of secure distributed computing. It is therefore highly desirable to provide 'provably secure' solutions, so that protocols for these goals can be raised above the previous (unsuccessful) attack-response design methodology. In this paper we propose protocols for the goals of entity authentication and authenticated key transport. It is demonstrated that these protocols provide provable security within a particularly powerful model of distributed computing. Roughly speaking, the process of proving security comes in five stages:

1. specification of model;
2. definition of goals within this model;
3. statement of assumptions;
4. description of protocol;
5. proof that protocol meets its goals within the model.

See [7] for an excellent account of the implications of provable security, and a precise description of the assurances that provable security offers. It should be noted that the usefulness of provable security rests fundamentally on the success with which the model mimics the actual environment in which the goal needs to be achieved. Further, it is highly desirable that any provably secure protocol should have comparable overheads to those used in practice, since anyone trying to implement solutions will usually favor those with lower overheads.

Some related work has examined the security of key distribution protocols through "zero-knowledge" proofs. However, this model closely mirrors only the prover-verifier world of the smart-card. Most key distribution protocols take place in the context of distributed computing, where a very different model is required[3]. It is precisely this environment that we address. Another approach, known as "authentication logic" and initiated in [9], analyses protocols in terms of the beliefs of the entities involved. Again, this is fundamentally different from the approach of this paper, in as much as such a logic can only ever provide a guarantee that a protocol does not succumb to the particular weaknesses captured by the logic being used. Our approach, on the other hand, enables us to say that a protocol cannot succumb to any attack in the model of distributed computing described.

Specifically the contributions of this paper are:

1. adapting the distributed computing model of Bellare and Rogaway [1] to the asymmetric case;

---

[3] Informally, 'distributed computing' refers to what we all think of as a computer network; that is a number of separate machines which can only communicate over public (unsecured) channels.

2. redefining security goals within this new model;
3. presenting and proving secure protocols for entity authentication and authenticated key transport within this model;
4. practical implementation of protocols at little security expense.

We particularly wish to stress the important role that an appropriate definition of the goal of a protocol plays in results of provable security—all protocols are provably secure under some definition; thus we believe that the emphasis in such work should be on how appropriate the model which admits provable security is, and on how appropriate the definition of the goal is, rather than the mere statement that such-and-such a protocol attains provable security. It is a central thesis of this work, therefore, that the model of distributed computing described faithfully models the environment in which solutions to the entity authentication and authenticated key transport problems are required, and that the formal definitions given for these goals are the 'right' ones.

2 Model of Distributed Environment

First some notation and language are introduced, and then the model itself is described. Since the model is an adaptation of the model used in [1], we give only a terse account here. A more complete explanation may be found in [1, 3].

2.1 Set-up

$\{0,1\}^*$ denotes the set of finite binary strings, and λ denotes the empty string. $I = \{1, 2, \ldots, N_1\}$ is the set of entities in this environment (the adversary is not included as an entity). We insist that $|I|$ be polynomial in the security parameter k, so that $N_1 = T_1(k)$ for some polynomial function T_1. $\mathbb{N}$ denotes the set of positive integers. A real-valued function $\epsilon(k)$ is *negligible* if for every $c > 0$ there exists $k_c > 0$ such that $\epsilon(k) < k^{-c}$ for all $k > k_c$.

Definition 1. A *protocol* is a pair $P = (\Pi, \mathcal{G})$ of probabilistic polytime computable functions (polytime in their first input):
Π specifies how (honest) players behave;
$\mathcal{G}$ generates key pairs for each entity.
The domain and range of these functions is as follows. Π takes as input:
1^k — the security parameter;
$i \in I$ — identity of sender;
$j \in I$ — identity of intended recipient;
$K_{i,j}$ — i's keying information K_i together with j's public keying information PK_j;
tran — a transcript of the protocol run so far (i.e. the ordered set of messages and appendices transmitted and received by i so far in this run of the protocol).
$\Pi(1^k, i, j, K_{i,j}, tran)$ outputs a triple $((m, a), \delta, \kappa)$, where:

$m \in \{0,1\}^* \cup \{*\}$ is the next message to be sent from i to j ($*$ indicates no message is sent);

a is an appendix to m (in the protocols described in this paper, a will always represent a signature on m);

$\delta \in \{\texttt{Accept}, \texttt{Reject}, *\}$ is i's current decision ($*$ indicates no decision yet reached);

κ is the exchanged key (which always has a null value unless the entity has accepted).

κ will not be used in protocols which achieve only entity authentication.

$\mathcal{G}$ takes as input the security parameter 1^k and generates independently and at random an authentication key pair, (PSK, SSK), for each entity using $\mathcal{G}_{sig}$, the key generation algorithm for a signature scheme of $\mathcal{G}$'s choice. In the case of key transport protocols, $\mathcal{G}$ also generates independently secrecy key pairs (PEK, SEK) using $\mathcal{G}_{enc}$, the key generation algorithm of a public-key encryption scheme of $\mathcal{G}$'s choice. These secrecy key pairs will be used by entities in key transport protocols to encrypt session keys. Finally, $\mathcal{G}$ forms a directory *public-info* consisting of a distinct triple corresponding to each entity — the triple corresponding to entity i consists of the identifier i, i's public signing key, and i's public encryption key.

$\mathcal{G}$ is a technical description of the key generation process. It is a formal model designed to capture the attributes of the techniques typically used to generate keys in a distributed environment. Of course, in real life, each entity will usually generate key pairs itself and then get them certified by a Certification Authority.

The keying information assigned to entity i by $\mathcal{G}$ will be denoted by $K_i = ((PSK_i, SSK_i), (PEK_i, SEK_i))$. K_i consists of two key pairs — the first pair is used by i for signing, and the second pair is used by i for encrypting. $PK_i = (PSK_i, PEK_i)$ represents i's public keying information, consisting of i's public signing key and public encryption key.

A generic execution of a protocol between two entities is called a *run* of the protocol. While a protocol is formally specified by a pair of functions $P = (\Pi, \mathcal{G})$, in this paper it is informally specified by the description of a run between two arbitrary entities. Any particular run of a protocol is called a *session*. The word 'session' is often associated with anything specific to one particular protocol run. For example, the keying information exchanged in the course of a protocol run is referred to as a *session key*. The individual messages that form a protocol run are called *flows*.

2.2 Description of Model

The adversary in this model is allowed enormous power. She controls all communication between entities, can at any time ask an entity to reveal any session key, or more seriously to reveal its long-term secret keys. Furthermore she may at any time initiate sessions between any two entities, engage the same two entities in simultaneous multiple sessions, and may ask an entity to enter a session with itself.

With such a powerful model it is not clear what it means for a protocol to be secure. Informally, we will say that an entity authentication protocol is secure if the only way the adversary can make uncorrupted entities (those whose long-term keying material she has not revealed) accept, is by relaying messages faithfully between them. In this case, the adversary effectively acts just like a wire.

We now formalize the above discussion.

An *adversary*, E, is a probabilistic polytime Turing Machine. E takes as input the security parameter, 1^k, along with *public-info*, the directory containing all entities' public keys. E has access to a collection of oracles:

$$\{\Pi_{i,j}^s : i \in I, j \in I, s \in \{1, 2, \ldots, N_2\}\} \ .$$

Oracle $\Pi_{i,j}^s$ behaves as entity i carrying out protocol Π in the belief that it is communicating with j for the sth time (i.e. the sth run of the protocol between i and j). Each $\Pi_{i,j}^s$ oracle maintains its own variable *tran* to store its view of the run so far. For each pair of entities (i, j), E is equipped with a polynomial number of oracles (so that $N_2 = T_2(k)$, where T_2 is a polynomial function). Each oracle $\Pi_{i,j}^s$ is initialized with the security parameter 1^k, entity i's keying information K_i, a *tran* value of λ, and the directory *public-info*.

E is allowed to make three types of queries of its oracles, as illustrated in the table below.

Query	Oracle reply	Oracle update
$\texttt{Send}(\Pi_{i,j}^s, (m', a'))$	$\Pi^{(m,a)\delta}(1^k, i, j, K_{i,j}, tran.(m', a'))$	$tran \leftarrow tran.(m', a').(m, a)$
$\texttt{Reveal}(\Pi_{i,j}^s)$	$\Pi^\kappa(1^k, i, j, K_{i,j}, tran)$	—
$\texttt{Corrupt}(i, K)$	K_i	$K_i \leftarrow K$

In the table, $\Pi^{(m,a)\delta}$ denotes the first two arguments of $\Pi_{i,j}$'s output, and Π^κ denotes the third. The $\texttt{Send}$ query represents E giving a particular oracle message m' with appendix a' as input. E initiates a session with the query $\texttt{Send}(\Pi_{i,j}^s, \lambda)$, i.e. by sending the oracle it wishes to start the session the empty string λ. $\texttt{Reveal}$ tells a particular oracle to reveal whatever session key it currently holds; this query is not relevant to protocols for entity authentication only. $\texttt{Corrupt}$ tells all $\Pi_{i,j}^s$ oracles, for any $j \in I$, $s \in \{1, 2, \ldots, N_2\}$, to reveal entity i's long-term secret keys to E, and further to replace K_i with any key pairs K of E's choice. In addition, all oracles' copies of i's public keying information in the directory *public-info* are updated. Additionally, E may use the $\texttt{Corrupt}$ query to merely update an entity's public keys. In this case E replaces the entity's private keys with λ, and its public keys with any public keys of E's choice. Now no oracle associated with the entity is able to compute any decryptions or signatures. This ability models the possibility that a real-life adversary may be able to get a public key certified without knowing the corresponding private key.

Our security definitions now take place in the context of the following experiment — the experiment of running a protocol $P = (\Pi, \mathcal{G})$ in the presence of an adversary E using security parameter k:

1. Toss coins for $\mathcal{G}$, E, and all oracles $\Pi_{i,j}^s$ (and if necessary for any random oracles employed by the signature scheme or encryption scheme);
2. Run $\mathcal{G}$ on input 1^k;
3. Initialize all $\Pi_{i,j}^s$ oracles;
4. Start E on input 1^k and *public-info*.

Now when E asks oracle $\Pi_{i,j}^s$ a query, $\Pi_{i,j}^s$ calculates the answer using the description of Π. This definition of the experiment associated with a protocol implies that when we speak of the probability that a particular event occurs during the experiment, then this probability is assessed over all the coin tosses made in step 1 above.

The first step in defining the security of a protocol is to show that the protocol is 'well-defined'. To assist in this process we sometimes need to consider the following particularly friendly adversary. For any pair of oracles $\Pi_{i,j}^s$ and $\Pi_{j,i}^t$, the *benign adversary* on $\Pi_{i,j}^s$ and $\Pi_{j,i}^t$ is the deterministic adversary that always performs a single run of the protocol between $\Pi_{i,j}^s$ and $\Pi_{j,i}^t$, faithfully relaying flows between these two oracles.

$tran_{i,j}^s$ will be used to denote the current state of $\Pi_{i,j}^s$'s variable *tran*. We say that $\Pi_{i,j}^s$ has *accepted* if

$$\Pi^\delta(1^k, i, j, K_{i,j}, tran_{i,j}^s) = \texttt{Accept} \ ,$$

it is *opened* if there has been a $\texttt{Reveal}(\Pi_{i,j}^s)$ query, and it is *corrupted* if there has been a $\texttt{Corrupt}(i, \cdot)$ query.

So far all we have done is describe the model. We are now ready to give formal definitions of the goals.

3   Entity Authentication

First, we'll look at entity authentication. The model described in §2 provides the necessary framework for our security proofs; however, before we can prove anything about any protocol, a formal definition of the goal of a secure entity authentication protocol must be given.

3.1   Definition of Security

The central idea is that of *matching conversations*. This was first formulated in [8], refined in [11], and later formalized in [1]. The concept of matching conversations has been slightly modified here, and *matching conversations including appendices* introduced, in order to cope with the adaptation of the model to asymmetric techniques. The reason for these changes lies in the definition of a secure signature scheme, and so a discussion is deferred until after Definition 5.

MATCHING CONVERSATIONS. Fix an execution of an adversary E. For any oracle $\Pi_{i,j}^s$ its *conversation* can be captured by a sequence $C = C_{i,j}^s =$

$$(\tau_1, (m_1, a_1), (\mu_1, \alpha_1)), (\tau_2, (m_2, a_2), (\mu_2, \alpha_2)), \ldots, (\tau_n, (m_n, a_n), (\mu_n, \alpha_n)) \ .$$

This sequence encodes that at time τ_1 oracle $\Pi_{i,j}^s$ was asked (m_1, a_1) and responded with (μ_1, α_1); and then at some later time $\tau_2 > \tau_1$, the oracle was asked (m_2, a_2) and answered (μ_2, α_2); and so forth, until finally, at time τ_n it was asked (m_n, a_n) and answered (μ_n, α_n). Adversary E terminates without asking $\Pi_{i,j}^s$ any more questions. If $\Pi_{i,j}^s$ has $(m_1, a_1) = \lambda$ it is called an *initiator oracle*; otherwise it is called a *responder oracle*.

We now define matching conversations and matching conversations including appendices. For simplicity we focus on the case where R, the number of flows, is odd. The case R even is analogous.

Definition 2. Fix a number of flows $R = 2\rho - 1$ and an R-flow protocol $P = (\Pi, \mathcal{G})$. Run P in the presence of an adversary E and consider two oracles $\Pi_{i,j}^s$, an initiator oracle, and $\Pi_{j,i}^t$, a responder oracle, that engage in conversations C and C' respectively.

1. C' is said to be a *matching conversation* to C if there exist $\tau_0 < \tau_1 < \cdots < \tau_{R-1}$ such that C is prefixed by:

$$(\tau_0, \lambda, (m_1, a_1)),\ (\tau_2, (\mu_1, \alpha_1'), (m_2, a_2)), \ldots, (\tau_{2\rho-2}, (\mu_{\rho-1}, \alpha_{\rho-1}'), (m_\rho, a_\rho))$$

and C' is prefixed by:

$$(\tau_1, (m_1, a_1'), (\mu_1, \alpha_1)),\ (\tau_3, (m_2, a_2'), (\mu_2, \alpha_2)), \ldots,$$
$$(\tau_{2\rho-3}, (m_{\rho-1}, a_{\rho-1}'), (\mu_{\rho-1}, \alpha_{\rho-1}))\ .$$

Further C' is a *matching conversation including appendices* to C if additionally $a_1 = a_1', \ldots, a_{\rho-1} = a_{\rho-1}'$ and $\alpha_1 = \alpha_1', \ldots, \alpha_{\rho-1} = \alpha_{\rho-1}'$.

2. C is said to be a *matching conversation* to C' if there exist $\tau_0 < \tau_1 < \cdots < \tau_R$ such that C' is prefixed by:

$$(\tau_1, (m_1, a_1'), (\mu_1, \alpha_1)),\ (\tau_3, (m_2, a_2'), (\mu_2, \alpha_2)), \ldots,$$
$$(\tau_{2\rho-3}, (m_{\rho-1}, a_{\rho-1}'), (\mu_{\rho-1}, \alpha_{\rho-1})),\ (\tau_{2\rho-1}, (m_\rho, a_\rho'), *)$$

and C is prefixed by:

$$(\tau_0, \lambda, (m_1, a_1)),\ (\tau_2, (\mu_1, \alpha_1'), (m_2, a_2)), \ldots, (\tau_{2\rho-2}, (\mu_{\rho-1}, \alpha_{\rho-1}',), (m_\rho, a_\rho))\ .$$

Further C is a *matching conversation including appendices* to C' if additionally $a_1 = a_1', \ldots, a_\rho = a_\rho'$ and $\alpha_1 = \alpha_1', \ldots, \alpha_{\rho-1} = \alpha_{\rho-1}'$.

Finally, $\Pi_{i,j}^s$ and $\Pi_{j,i}^t$ are said to have had *matching conversations including appendices*, if C is a matching conversation including appendices to C' and C' is a matching conversation including appendices to C.

Roughly speaking, this definition of matching conversations captures when E faithfully relays messages. In the first case, E relays all $\Pi_{i,j}^s$'s messages (except possibly the last) to $\Pi_{j,i}^t$, and then relays the replies back. The second case implies the first case, but in addition $\Pi_{i,j}^s$'s last message is relayed to $\Pi_{j,i}^t$.

Matching conversations including appendices mean that E also relays appendices faithfully between the two oracles.

AUTHENTICATION. We are now ready to define a secure authentication protocol. The definition adopted is an adaptation of the definition for the symmetric case introduced in [1]. Let $\texttt{No-Matching}^E(k)$ denote the event that, when protocol P is run against adversary E, there exists an oracle $\Pi_{i,j}^s$ with $i, j \notin C$ (where C denotes the set of entities corrupted by E) which accepted but there is no oracle $\Pi_{j,i}^t$ which has had a matching conversation to $\Pi_{i,j}^s$.

Definition 3. A protocol P is a *secure (mutual) authentication protocol* if for every adversary E:

1. If $\Pi_{i,j}^s$ and $\Pi_{j,i}^t$ have matching conversations including appendices, then both oracles accept;
2. The probability of $\texttt{No-Matching}^E(k)$ is negligible.

The first condition says that if both entities behave honestly, and the transmissions between them are not tampered with, then both accept. The second says that essentially the only way for any adversary to get an uncorrupted oracle $\Pi_{i,j}^s$ to accept in a run of the protocol with any other uncorrupted entity is to pick an oracle $\Pi_{j,i}^t$ and relay messages faithfully between $\Pi_{i,j}^s$ and $\Pi_{j,i}^t$.

Note that although our protocols will employ a signature scheme, no concept of non-repudiation is included in the above definition. We believe that non-repudiation is a secondary issue in entity authentication and key transport. Of course, some applications may desire non-repudiation, possibly motivating future examination of a goal like *authenticated key transport with non-repudiation of session keys*.

3.2 Protocol

We are now ready to specify our entity authentication protocol. The cryptographic primitive needed is a secure signature scheme.

SIGNATURE SCHEMES. Provably secure signature schemes were first discussed by Goldwasser, Micali, and Rivest [14]. Since then, efficient provably secure schemes have been introduced [2, 12, 6, 21].

Definition 4 [14]. A *signature scheme* is a triple $(\mathcal{G}_{sig}, [\cdot]_{sig_{(\cdot)}}, [\cdot, \cdot]_{ver_{(\cdot)}})$ of poly-time algorithms. On input 1^k, $\mathcal{G}_{sig}$ generates a key pair (PSK, SSK). To sign a message m, the entity with key pair (PSK, SSK) computes:

$$(m, \sigma) \leftarrow [m]_{sig_{SSK}} \; .$$

The *signature* for m is σ, and (m, σ) is the *signed message*. To verify (m, σ), compute:

$$[m, \sigma]_{ver_{PSK}} \in \{\texttt{True}, \texttt{False}\} \; .$$

We demand that:

$$[m, \sigma]_{ver_{PSK}} = \texttt{True} \text{ for all } (m, \sigma) \in \{[m]_{sig_{SSK}}\} \; .$$

The key generation and the signature algorithms are allowed to be probabilistic, while the verification algorithm is deterministic.

An *adversary* F (of just the signature scheme) is a probabilistic polytime algorithm with access to a signing oracle (and, if appropriate, a public random oracle). The input to F is a public key PSK chosen by running $\mathcal{G}_{sig}$, and the output of F is a pair (m, σ) such that F did not query its signing oracle on m.

Definition 5 [14]. A signature scheme is a *secure signature scheme* if for every adversary F (of just the signature scheme), the function $\epsilon(k)$ defined by:

$$\epsilon(k) = Pr[(PSK, SSK) \leftarrow \mathcal{G}_{sig}(1^k); (m, \sigma) \leftarrow F(PSK) : [m, \sigma]_{ver_{PSK}} = \textbf{True}]$$

is negligible. (In the random oracle model, the above probability is also assessed over the coin tosses of a random oracle.)

Roughly speaking, this means a signature scheme is secure only if the probability of forging a signature on a new message is negligible. Thus we require a secure signature scheme to withstand an adaptive chosen-message attack.

COMMENTS. Let us now return to discuss the slight modification to the definition of matching conversations. Under the standard definition of matching conversations [1], C and C' are matching conversations if both messages and appendices are relayed faithfully by E. However, in the definition of secure signature scheme, an adversary F is not excluded from querying its signing oracle on m, receiving (m, σ) as the reply, then computing $\sigma' \neq \sigma$ such that σ' is also a valid signature. This ability makes the definition of secure signature scheme incompatible with the standard definition of matching conversations. For this reason, the definition of matching conversations has been updated so that the standard definition of a secure signature scheme can be used but the definition of security for protocols is not compromised. (Note that an accepting oracle $\Pi_{i,j}^t$ is still assured that there is a $\Pi_{j,i}^t$ oracle with whom *messages* have been relayed back and forth in real time.)

While this solution appears ideal — allowing the standard definition of a secure signature scheme to be used without compromising protocol security, let us briefly describe how the definition of secure signature scheme could be modified so that the standard definition of matching conversations could be used. Instead of having F (the adversary of the signature scheme) output a pair (m, σ) such that F did not query its signing oracle on m, F must now output a pair (m, σ) such that F's signing oracle did not answer one of F's queries with (m, σ). This modification would allow the standard definition of matching conversations to be used in the definition of protocol security (note that the standard definition of matching conversations corresponds to our definition of matching conversations including appendices). Further, call a signature scheme *deterministic* if its signing algorithm is deterministic. Then any deterministic signature scheme secure under Definition 5 (e.g., FDH in [6]) is also secure under the modified definition (since then each m always has the same signature σ). It would be interesting to investigate whether the known provably secure probabilistic signature schemes (e.g., PSS in [6]) also satisfy the stronger definition of security outlined above.

If the random oracle model is being used for the signature scheme concerned, then our model of distributed computing should be updated so that all oracles and E are themselves supplied with a public random oracle, $\mathcal{H}_{sig}$ (for use with the signature scheme). The security proofs remain unaffected by adding to the model in this way.

AUTHENTICATION PROTOCOL. We can now finally describe the protocol. It is represented graphically in Figure 1. Use $\in_R$ to denote an element chosen independently at random, and commas to denote a unique encoding through concatenation (or any other unique encoding). When entity i wishes to initiate a run of Π with entity j, i selects $R_i \in_R \{0,1\}^k$, and sends the message R_i to j (with no appendix). On receipt of this string, j selects $R_j \in_R \{0,1\}^k$, computes the signature on the string i, R_j, R_i using the private signing key SSK_j, and sends this signed message to i. (Here, i, R_j, R_i form the message, and its signature σ the appendix.) i checks that the identity i in the message is as intended, and the same R_i that i generated in flow 1 is present, and if so verifies j's signature using the distributed copy of PSK_j. If the signature is valid, then i in turn signs the string j, R_j using the private signing key SSK_i. i sends this signed message to j and accepts. (As before, j, R_j forms the message, and its signature σ the appendix.) j checks the correct j, R_j are present, and if so verifies the signature. If the signature is valid, j accepts. If at any stage, a signature fails to validate or one of the flows is not in the correct form, then that party terminates the protocol run, and rejects.

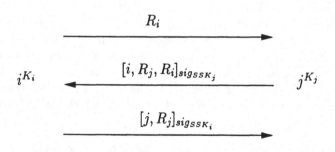

Fig. 1. Protocol 1

Theorem 6. *Protocol 1 is a secure (mutual) authentication protocol provided the signature scheme is secure.*

The proof of this theorem is presented in Appendix A.

COMMENTS. Notice that the third flow of Protocol 1 includes the identity of j, in contrast to Bellare and Rogaway's protocol MAP1 [1] where the identity of i is included. This change is necessary due to the adaptation of our model to incorporate corruption of entities together with the use of asymmetric techniques. In the symmetric setting, i's final message is unique to entities sharing key a. In the asymmetric setting, merely signing R_j using i's secret information is insufficient, since i signs messages in the same way no matter which entity it believes it is communicating with.

We remark in passing that it is easy to show that the probability that an uncorrupted oracle has a matching conversation with two other uncorrupted oracles in a run of Protocol 1 is negligible. Additionally, we mention that both i and j can optionally add extra data to the signed strings to flows 2 and 3. Such a variant will be used in the next section to transport an encrypted key. Adding data to the signatures in this way does not affect the proof of security, although care may be required to ensure that the encodings remain unique.

4 Authenticated Key Transport

We now extend our protocol to transfer session keys. Since 'distributed networks' are the typical setting for the authenticated key transport problem as well as the entity authentication problem, the model remains the same. However more definitions are needed: what does it mean for an authenticated key transport protocol to be secure?

4.1 Definition of Security

Fix $S = \{S_k\}_{k \in \mathbb{N}}$ with each S_k a samplable distribution over $\{0, 1\}^{h(k)}$ for some polynomial function h. The intent of an authenticated key transport protocol is to authenticate entities and to securely transfer a key sampled from S_k. We stipulate that an oracle never holds a session key κ unless it has accepted.

In this context the **Reveal** query captures the notion that compromising other session keys should have no effect on the security of the current run. An oracle $\Pi_{i,j}^s$ is said to be *fresh* if it has accepted, neither i nor j has been corrupted, it remains unopened, and there is no opened oracle $\Pi_{j,i}^t$ which has had a matching conversation to $\Pi_{i,j}^s$.

KEY TRANSPORT. We want the adversary to be able to learn as little as possible about the session key of any fresh oracle. This is formalized along the lines of polynomial indistinguishability. Specifically at the end of its run, the adversary should be unable to gain more than a negligible advantage over simply guessing when it tries to distinguish the actual key held by a fresh oracle from a key sampled at random from S_k.

Therefore the following addendum is made to the experiment. After the adversary has asked all the queries it wishes to make, the adversary asks a fresh oracle $\Pi_{i,j}^s$ a single new query:

$$\mathbf{Test}(\Pi_{i,j}^s) \ .$$

To answer the query, the oracle flips a fair coin $b \leftarrow \{0,1\}$, and returns the session key $\kappa_{i,j}^{s}$ if $b = 0$, or else an independent random sample from S_k if $b = 1$. The adversary's job is now to guess b. To this end, E outputs a bit Guess. Let Good-Guess$^{E}(k)$ be the event that Guess $= b$. Then we define:

$$advantage^{E}(k) = |Pr[\text{Good-Guess}^{E}(k)] - \tfrac{1}{2}| .$$

Definition 7. A protocol $P = (\Pi, \mathcal{G})$ is a *secure authenticated key transport protocol* if P is a secure (mutual) authentication protocol and:

1. If E is the benign adversary on $\Pi_{i,j}^{s}$ and $\Pi_{j,i}^{t}$, then both oracles always accept holding the same session key κ, and this variable is distributed according to S_k;
2. For any adversary E, $advantage^{E}(k)$ is negligible.

This says that, in addition to being a secure entity authentication protocol, we require that oracles who have matching conversations including appendices in the presence of a benign adversary hold the same session key, which is correctly distributed, and that no adversary can deduce any information about the session key of a fresh oracle. Definition 7 is analogous to the definition given for authenticated key transport in the symmetric setting in [1].

4.2 Protocols

Now we'll specify our authenticated key transport protocol. The cryptographic primitive required is a secure public-key encryption scheme.

ENCRYPTION SCHEMES. Provably secure encryption schemes were introduced in [13]. The variants we need withstand adaptive chosen-ciphertext attacks. These were first discussed in [22]. Efficient provably secure schemes in the 'random oracle model' are demonstrated in [2, 4].

Definition 8 [22, 2]. An *encryption scheme* is a triple $(\mathcal{G}_{enc}, [\cdot]_{enc_{(.)}}, [\cdot]_{dec_{(.)}})$ of polytime algorithms. On input 1^k, $\mathcal{G}_{enc}$ generates an encryption key pair (PEK, SEK). To encrypt a message m to send to the entity with key pair (PEK, SEK), compute and send:

$$c \leftarrow [m]_{enc_{PEK}} .$$

To decrypt c, the entity computes:

$$[c]_{dec_{SEK}} .$$

We demand that:

$$[[m]_{enc_{PEK}}]_{dec_{SEK}} = m \text{ for all } m .$$

An *adversary* F (of the encryption scheme) is a pair of probabilistic polytime algorithms (A_1, A_2) with access to a decryption oracle (and if appropriate a public random oracle) enabling the adversary to mount an adaptive chosen-ciphertext attack on the scheme. The input to F is a public key PEK selected using $\mathcal{G}_{enc}$. A_1 selects a pair of messages, m_0 and m_1, and A_2 is given α, an encryption of either m_0 or m_1. A_2 must guess whether α is an encryption of m_0 or m_1. We stipulate that A_2 does not query its decryption oracle with α. The security of an encryption scheme is now defined in terms of polynomial indistinguishability:

Definition 9 [22, 2]. An encryption scheme is a *secure encryption scheme* if for every adversary F (of just the encryption scheme), the function $\epsilon(k)$ defined by:

$$\epsilon(k) = |Pr[(PEK, SEK) \leftarrow \mathcal{G}_{enc}(1^k); (m_0, m_1) \leftarrow A_1(PEK);$$
$$b \leftarrow \{0, 1\}; \alpha \leftarrow [m_b]_{enc_{PEK}} : A_2(PEK, m_0, m_1, \alpha) = b] - \tfrac{1}{2}|$$

is negligible. (In the random oracle model, the above probability is additionally assessed over the random oracle's coin tosses.)

Roughly speaking, this means an encryption scheme is secure only if the advantage gained by any adversary in guessing which of two messages of the adversary's choice is encrypted, through seeing one of their encryptions, is negligible, even after an adaptive chosen-ciphertext attack.

As before, if the random oracle paradigm is being used for the encryption scheme, then E and all oracles in our model must be supplied with a public random oracle, $\mathcal{H}_{enc}$, for use when encrypting and decrypting. The presence of this additional oracle does not affect the security proofs.

KEY TRANSPORT PROTOCOL. Our protocol is represented in Figure 2. The actions carried out by entities i and j are similar to the actions of Protocol 1, with the following additions. In reply to flow 1, j also selects $\kappa \in_R S_k$, and encrypts j, κ using the public encryption key PEK_i. j then adds this ciphertext to the string that it signs. On receiving flow 2, after checking that the signature it received is valid, i additionally decrypts $[j, \kappa]_{enc_{PEK_i}}$, checks that j and κ are valid, then accepts with κ as the exchanged session key. Once j has accepted, j also accepts κ as the session key. As before, if any of the checks performed by an entity fail, then that entity rejects and terminates the protocol run.

Recall that SEK_i used in the encryption function was chosen by $\mathcal{G}$ independently from SSK_i used in the signature function.

Theorem 10. *Protocol 2 is a secure authenticated key transport protocol provided that the encryption and signature schemes are secure.*

The proof of this theorem is presented in Appendix B.

COMMENTS. Including j in the encrypted string is required to avoid the following attack using the Corrupt query. E corrupts any other entity, then enters into

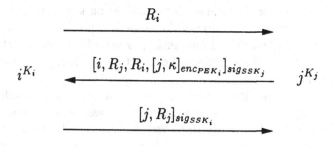

$$R_i$$

$$_i{}^{K_i} \qquad [i, R_j, R_i, [j, \kappa]_{enc_{PEK_i}}]_{sig_{SSK_j}} \qquad _j{}^{K_j}$$

$$[j, R_j]_{sig_{SSK_i}}$$

Fig. 2. Protocol 2

a run of the protocol with i, behaving as the corrupted entity. E behaves as specified by P, except that, instead of sampling a key, she sends an encrypted string that she has observed during a bona fide run of the protocol between i and j. E can now discover the session key from this bona fide run using the **Reveal** query.

5 Practicalities

'REAL WORLD' IMPLICATIONS. What are the implications of these theoretical results above to the 'real world'?

Prior to the work of Bellare and Rogaway, most results on provable security did not lead to practical implementations. As in [1, 2, 4, 5], this is not the case here. In fact, both Protocols 1 and 2 are marginally more efficient than those standardized by ISO [15, 16] and NIST [20], thus the overheads involved in implementing these protocols are almost identical to those of protocols in widespread use today. In fact, our results imply that the three-pass protocols standardized in [15, 16, 20] are provably secure in the model we propose when the signature and encryption schemes employed are themselves provably secure.

Implementors who are prepared to accept the so-called 'random oracle paradigm' [2] can implement provably secure encryption and signature schemes that are as efficient as any used in practice [4, 6, 7, 21, 23]. Implementors who are not prepared to accept this assumption must accept some additional overheads if they wish to implement the protocols precisely [14, 22, 12].

However, we believe that the most significant practical implication of our results is that they offer some degree of security even if the protocols are implemented with encryption and signature schemes that do not provide provable security. In this situation, the implementor is assured that any attack on the implementation would have to exploit a weakness in the signature scheme or encryption scheme — so that the protocols avoid *generic* attacks (those that do not exploit properties of the specific signature scheme or encryption scheme). Typically, attacks on encryption schemes and signature schemes have a high computational overhead, while attacks on previous key transport protocols have virtually no computational overhead. Most attacks on key transport protocols exploit some subtle flaw in the form of the flows: it is precisely such a weakness that our result prohibits. Indeed the authors know of no serious attack on a key transport protocol which exploits the structure of the signature scheme or encryption scheme used.

Thus when employing Protocols 1 or 2 even without using provably secure encryption and signature schemes, an implementor is assured that no subtle flaws exist in the protocol. No currently employed protocol can give such a guarantee and, as discussed in §1, a large number of flawed protocols have previously been proposed. Thus we believe that our results fulfill a 'real need' in the cryptographic world.

IMPLEMENTATION ISSUES. When implementing these protocols, bandwidth savings can easily be made in flows 2 and 3. For example, in Protocol 1, the signatures sent in these flows contain some redundancy, so that in flow 2, instead of sending the pair (m, σ) where $m = (i, R_j, R_i)$, j only needs to send (R_j, σ), since the rest of the flow can be inferred by i. Similarly, in flow 3, instead of sending (m, σ) where $m = (j, R_j)$, just sending σ is sufficient.

Finally, we note that since E may use her **Corrupt** query to get public keys certified without knowing the corresponding private keys, E may be able to pass off strings signed by other oracles as her own. It is clear that E gains no advantage on the authentication protocol or authenticated key transport protocol by doing so; however if the signed string includes some additional data, such a property may be undesirable. If the particular application requires it, this ability can be removed by including the identity of j in flow 2, and the identity of i in flow 3. Alternatively, when an entity attempts to get a public key certified, the Certification Authority can check knowledge of the corresponding private key. We believe that this is, in any case, a sensible precaution in any implementation of a Certification Hierarchy, particularly if any form of non-repudiation is required.

6 Conclusions and Further Work

We have proposed protocols for entity authentication and authenticated key transport, and provided proofs that the protocols meet their security goals. In particular, this means that the protocols prevent both passive and active attacks, including so-called known key attacks, in which an adversary is able to gain previous session keys. We have supplied strong evidence that practical implementations of the protocols also offer superior security assurances than other protocols currently in use.

The authors believe there are a number of aspects of this work that require further discussion. In particular, questions arising are: is the model of distributed computing ideal? What impact do security proofs have on protocols? Can these methods be applied to protocols with different security goals? For which other goals would implementors like to see proven secure solutions?

Finally, the authors note that it is possible to use these results to obtain exact security [6]. By following the proofs carefully, it is possible to make specific recommendation about, for example, the length of random challenges that should be used, or the ideal security level that the signature scheme should attain, given a desired security assurance from the protocol itself.

7 Acknowledgements

The authors wish to thank Don Johnson, Chris Mitchell, and Peter Wild for helpful comments on an earlier draft of this work, and Phil Rogaway for an enlightening conversation at PKS'97.

References

1. M. Bellare and P. Rogaway. Entity authentication and key distribution. In *Advances in Cryptology: Crypto '93*, pages 232–249, 1993.
2. M. Bellare and P. Rogaway. Random oracles are practical: a paradigm for designing efficient protocols. In *1st ACM Conference on Computer and Communications Security*, pages 62–73, 1993.
3. M. Bellare and P. Rogaway. Entity authentication and key distribution. Full version of [1], available at http://www-cse.ucsd.edu/users/mihir.
4. M. Bellare and P. Rogaway. Optimal asymmetric encryption. In *Advances in Cryptology: Eurocrypt '94*, pages 92–111, 1995.
5. M. Bellare and P. Rogaway. Provably secure session key distribution—the three party case. In *Proceedings of the 27th ACM Symposium on the Theory of Computing*, pages 57–66, 1995.
6. M. Bellare and P. Rogaway. The exact security of digital signatures—how to sign with RSA and Rabin. In *Advances in Cryptology: Eurocrypt '96*, pages 399–416, 1996.
7. M. Bellare and P. Rogaway. Minimizing the use of random oracles in authenticated encryption schemes. In *Proceedings of PKS'97*, 1997.
8. R. Bird, I. Gopal, A. Herzberg, P. Janson, S. Kutten, R. Molva, and M. Yung. Systematic design of two-party authentication protocols. In *Advances in Cryptology: Crypto '91*, pages 44–61, 1991.
9. M. Burrows, M. Abadi, and R. Needham. A logic of authentication. DEC SRC report 39, Digital Equipment Corporation, Palo Alto, CA, Feb. 1989. Revised Feb. 1990.
10. W. Diffie and M. Hellman. New directions in Cryptography. *IEEE Transactions on Information Theory*, IT-22(6): 644–654, November 1976.
11. W. Diffie, P.C. van Oorschot, and M.J. Wiener. Authentication and authenticated key exchanges. *Designs, Codes, and Cryptography*, 2: 107–125, 1992.
12. C. Dwork and M. Naor. An efficient existentially unforgeable signature scheme and its applications. In *Advances in Cryptology: Crypto '94*, pages 234–246, 1994.
13. S. Goldwasser and S. Micali. Probabilistic encryption. *Journal of Computer and System Sciences*, 28: 270–299, 1984.
14. S. Goldwasser, S. Micali, and R. Rivest. A digital signature scheme secure against adaptive chosen message attacks. *SIAM Journal of Computing*, 17(2): 281–308, 1988.

15. ISO/IEC 9798-3. *Information technology – Security techniques – Entity authentication mechanisms – Part 3: Entity authentication using a public-key algorithm*, International Organization for Standardization, Geneva, Switzerland, 1993 (first edition).

16. ISO/IEC 11770-3. *Information technology – Security techniques – Key management – Part 3: Mechanisms using asymmetric techniques*, draft, (DIS), 1996.

17. A.J. Menezes, P.C. van Oorschot, and S.A. Vanstone. *Handbook of Applied Cryptography*, chapter 12. CRC Press, 1996.

18. R.C. Merkle. Secure communications over insecure channels. *Communications of the ACM*, 21: 294–299, 1978.

19. J.H. Moore. Protocol failure in cryptosystems. Chapter 11 in *Contemporary Cryptology: the Science of Information Integrity*, G. J. Simmons, editor, 541–558, IEEE Press, 1992.

20. National Institute of Standards and Technology, *Entity Authentication using Public Key Cryptography*, FIPS 196, February, 1997.

21. D. Pointcheval and J. Stern. Security proofs for signature schemes. In *Advances in Cryptology: Eurocrypt '96*, pages 387–398, 1996.

22. C. Rackoff and D.R. Simon. Non-interactive zero-knowledge proof of knowledge and chosen ciphertext attack. In *Advances in Cryptology: Crypto '91*, pages 433–444, 1992.

23. Y. Zheng and J. Seberry. Immunizing public key cryptosystems against chosen ciphertext attacks. *IEEE Journal on Selected Areas in Communications*, 11(5): 715–724, 1993.

A Proof of Theorem 6

Theorem 6 *Protocol 1 is a secure (mutual) authentication protocol provided the signature scheme is secure.*

Proof. The first condition of Definition 3, namely that two honest parties both accept if the adversary acts as a wire, is clear. Therefore we concentrate on the second condition.

Carry out the experiment against an arbitrary adversary E. Suppose, by way of contradiction, that No-Matching$^E(k)$ is non-negligible. We shall say that E *succeeds* if, at the end of E's operation, there exists an oracle $\Pi_{i,j}^s$ with $i,j \notin C$ which has accepted but no oracle $\Pi_{j,i}^t$ has had a matching conversation to $\Pi_{i,j}^s$. Further in this case we say that E has succeeded *against* $\Pi_{i,j}^s$. Hence, by assumption:

$$Pr[E \text{ succeeds}] = n(k)$$

with $n(k)$ non-negligible. Let $n_1(k)$ denote the probability E succeeds against at least one initiator oracle, and $n_2(k)$ the probability E succeeds against at least one responder oracle but no initiator oracles. Then

$$n(k) = n_1(k) + n_2(k) \ .$$

We conclude that at least one of $n_1(k)$ or $n_2(k)$ must be non-negligible. Therefore the examination splits into two cases.

Case 1. Suppose $n_1(k)$ is non-negligible, i.e. E is successful against an initiator oracle with non-negligible probability. In this case we construct from E an adversary F of the signature scheme which is successful with non-negligible probability.

Recall that F's experiment begins as follows. Coins are tossed for F, $\mathcal{G}_{sig}$, and F's signing oracle (and if necessary $\mathcal{H}_{sig}$), then $\mathcal{G}_{sig}$ is run on input 1^k, producing a key pair (PSK, SSK). F is started on input PSK.

Now F must perform E's experiment. F picks at random an entity $j \in I$, guessing that E will succeed against an initiator $\Pi_{i,j}^s$ oracle for some i and s. F tosses coins as required, and instead of running $\mathcal{G}$, F chooses entities' key pairs itself. F selects all key pairs except j's by running $\mathcal{G}_{sig}$. It chooses its own input, PSK, as j's public signing key. F forms the directory *public-info* and starts E. F answers all E's oracle queries itself as specified by Π, using its own signing oracle to help answer queries for entity j.

Suppose E does succeed against some initiator $\Pi_{i,j}^s$. Then at some time τ_0, E queries $\Pi_{i,j}^s$ with λ, and F replies on behalf of $\Pi_{i,j}^s$ with R_i. At some time $\tau_2 > \tau_0$, $\Pi_{i,j}^s$ must receive

$$[i, R_j, R_i]_{sig_{SSK}}$$

for some R_j. If a signature on this message has not previously been output by F on behalf of entity j, then F can use the flow to win its own experiment. On the other hand, if F has previously computed $[i, R_j, R_i]_{sig_{SSK}}$ on behalf of entity j, then the form of the flow implies that it was on behalf of a responder $\Pi_{j,i}^t$ oracle which received R_i as its own first flow. If this happened at time $\tau_1 < \tau_0$, then F gives up — however the probability this happens is negligible since R_i was picked at random. It cannot have happened at time $\tau_1 > \tau_0$, since by assumption $\Pi_{j,i}^t$ has not had a matching conversation to $\Pi_{i,j}^s$.

If E does not succeed against some initiator $\Pi_{i,j}^s$ oracle, then F gives up.

In this way F can use E to win its experiment with probability at least

$$\frac{n_1(k)}{T_1(k)}(1 - \lambda(k))$$

for some negligible $\lambda(k)$ — this is still non-negligible, contradicting the assumed security of the signature scheme. We conclude that $n_1(k)$ is negligible.

Case 2. Suppose $n_2(k)$ is non-negligible, i.e. E is successful against a responder oracle but no initiator oracles with non-negligible probability. As before, we use E to construct an adversary F of the signature scheme.

F's actions are similar to those of the adversary constructed in Case 1. F picks at random $i \in I$, guessing that E will succeed against a responder $\Pi_{j,i}^t$ oracle for some j and t, but won't succeed against any initiator oracles. Again, F selects all key pairs except i's itself using $\mathcal{G}_{sig}$, and chooses its own input, PSK, as i's public signing key. F forms the directory *public-info* itself and starts E. F

answers all E's oracle queries as specified by Π, using its own signing oracle to perform signatures on behalf of entity i.

Suppose E does succeed against some responder $\Pi_{j,i}^t$, and E is not successful against any initiator oracles. Then at some time τ_1, $\Pi_{j,i}^t$ received flow R_i and responded with

$$[i, R_j, R_i]_{sig_{SSK_j}} .$$

It must at some time $\tau_3 > \tau_1$ receive $[j, R_j]_{sig_{SSK}}$. If F has not previously computed a signature on this message on behalf of entity i, then F can use the flow to win its own experiment. On the other hand, if F has previously signed this message on behalf of i, its form implies it must have been as the final flow of a $\Pi_{i,j}^s$ oracle. The interaction of such a $\Pi_{i,j}^s$ oracle with E in general has the form:

$$(\tau_0, \lambda, R_i'), \quad (\tau_2, [i, R_j', R_i']_{sig_{SSK_j}}, [j, R_j']_{sig_{SSK}})$$

for some $\tau_2 > \tau_0$. Such a $\Pi_{i,j}^s$ oracle must have accepted, so by assumption some $\Pi_{j,i}^u$ has had a matching conversation with $\Pi_{i,j}^s$. If $u \neq t$, then F gives up — however the probability of this event is negligible, since then $R_j' = R_j$ with R_j' and R_j chosen independently at random. The case $u = t$ is excluded, since in that case $\tau_0 < \tau_1 < \tau_2 < \tau_3$, and $\Pi_{i,j}^s$ has a matching conversation to $\Pi_{j,i}^t$.

If E does not succeed against some responder $\Pi_{j,i}^t$ oracle, or if E succeeds against an initiator oracle, then F gives up.

The F constructed uses E to win its experiment with probability at least:

$$\frac{n_2(k)}{T_1(k)}(1 - \lambda(k))$$

for some negligible $\lambda(k)$ — again this is still non-negligible, contradicting the assumed security of the signature scheme. We conclude that $n_2(k)$ is also negligible.

Together the two cases contradict the assumption that $n(k)$ is non-negligible. We conclude that $\text{No-Matching}^E(k)$ is negligible for all adversaries E. $\square$

B Proof of Theorem 10

Theorem 10 *Protocol 2 is a secure authenticated key transport protocol provided that the encryption and signature schemes are secure.*

Proof. P is a secure (mutual) authentication protocol, since it is identical to the variant of Protocol 1 with data optionally included in flows 2 and 3.

The first condition of Definition 7 is also clearly satisfied; that is, in the presence of a benign adversary that passes flows between $\Pi_{i,j}^s$ and $\Pi_{j,i}^t$, these oracles will always accept holding the same key, and this key will be distributed according to S_k. It remains to prove the second condition, namely that $advantage^E(k)$ is negligible for all adversaries E. We argue by contradiction. Run the experiment against an arbitrary adversary E, and suppose that $advantage^E(k)$ is non-negligible.

Now we say that E *succeeds* (against $\Pi^s_{i,j}$) if E picks $\Pi^s_{i,j}$ to ask its **Test** query and outputs the correct bit **Guess**. Let $\frac{1}{2} + n(k)$ represent the probability that E succeeds, so that $n(k)$ is, by assumption, non-negligible. Let B_k be the event that E picks an initiator oracle to ask its **Test** query, so that $\overline{B}_k$ is the event that E picks a responder oracle. Further let $\frac{1}{2} + n_1(k)$ be the probability that E succeeds having picked an initiator oracle, and let $\frac{1}{2} + n_2(k)$ be the probability that E succeeds having picked a responder oracle. Then we have:

$$Pr[E \text{ succeeds}] = Pr[E \text{ succeeds}|B_k]Pr[B_k] + Pr[E \text{ succeeds}|\overline{B}_k]Pr[\overline{B}_k] \ ,$$

so that $n(k) = n_1(k)Pr[B_k] + n_2(k)Pr[\overline{B}_k]$. We conclude that at least one of the following must be true: $n_1(k)$ and $Pr[B_k]$ are non-negligible, or $n_2(k)$ and $Pr[\overline{B}_k]$ are non-negligible. Therefore the examination splits into two cases:

<u>Case 1.</u> Suppose $n_2(k)$ and $Pr[\overline{B}_k] = n_4(k)$ are non-negligible, i.e. E picks a responder oracle with non-negligible probability and in this case succeeds with probability exceeding $\frac{1}{2}$ by a non-negligible amount.

We construct from E an adversary $F = (A_1, A_2)$ of the encryption scheme. Recall F's experiment (see Definition 9). Coins are tossed for F, $\mathcal{G}_{enc}$, and F's decryption oracle (and possibly a random oracle $\mathcal{H}_{enc}$). Then $\mathcal{G}_{enc}$ is run on input 1^k, producing a key pair (PEK, SEK). F itself takes as input PEK. A_1 now computes using F's oracles, and outputs two messages m_0 and m_1. In reply to A_1's output, one of m_0 and m_1 is picked at random, and it is encrypted under SEK, giving α. A_2 now takes as input PEK, m_0, m_1, and α, and must try to guess whether α is the encryption of m_0 or m_1. Note that A_2 is forbidden from calling its decryption oracle on α.

The $F = (A_1, A_2)$ adversary we construct using E works as follows. A_1 selects at random an entity $j \in I$. A_1 makes two independent samples from S_k, κ_0 and κ_1. Then A_1 outputs $m_0 = (j, \kappa_0)$ and $m_1 = (j, \kappa_1)$ and halts.

A_2 gets as input PEK, (j, κ_0), (j, κ_1), and α, the encryption of either (j, κ_0) or (j, κ_1) (and must try to guess which one). To this end, A_2 picks at random $i \in I$ and $t \in \{1, \ldots, N_2\}$, guessing that E will select responder oracle $\Pi^t_{j,i}$ which has had a matching conversation with some initiator oracle to ask its **Test** query. A_2 now performs E's experiment as follows. A_2 tosses coins for $\mathcal{G}_{sig}$, $\mathcal{G}_{enc}$, E, $\Pi^s_{i,j}$ oracles, and any random oracles as required. Instead of running $\mathcal{G}$, A_2 picks all entities' keys itself independently, using $\mathcal{G}_{sig}$ and $\mathcal{G}_{enc}$, except for the encryption key of i. It chooses F's input PEK to be i's public encryption key.

A_2 now starts E, and answers all E's queries itself according to Π (employing its own oracles as necessary to help answer **Reveal** queries that E asks entity i and to decide whether or not i's oracles should accept) except for the first **Send** query to $\Pi^t_{j,i}$ and any queries to entity i that require decryption of α.

A_2 answers E's first **Send** query to $\Pi^t_{j,i}$ as follows. If E does not invoke $\Pi^t_{j,i}$ as a responder oracle, A_2 gives up and guesses at random whether α is the encryption of m_0 or m_1. If E invokes $\Pi^t_{j,i}$ as a responder oracle, then when E

sends R_i to $\Pi_{j,i}^t$, A_2 replies on behalf of $\Pi_{j,i}^t$ with

$$[i, R_j, R_i, \alpha]_{sig_{SSK_j}}$$

for some $R_j \in_R \{0,1\}^k$.

If A_2 ever needs to decrypt α (on behalf of i) to answer one of E's queries then A_2 gives up and guesses at random — notice however that such a call can only be required for an initiator $\Pi_{i,j}$ oracle that needs to decrypt α to answer a **Reveal** query. For this decryption call cannot have been required on behalf of a $\Pi_{i,l}$ oracle with $l \neq j$ — if such an oracle receives the encrypted string α, A_2 knows the string does not decrypt to the correct form (l, κ), so has the oracle reject. Furthermore, when a $\Pi_{i,j}$ oracle receives α in this way, A_2 knows the oracle should accept, since the string certainly decrypts to the correct form (j, κ) for some κ.

Suppose that E does pick responder $\Pi_{j,i}^t$ which has had a matching conversation with some $\Pi_{i,j}^s$ to ask its **Test** query. Suppose A_2 has not needed to call its own decryption oracle on α to answer E's queries. Then A_2 can use E's bit **Guess** to predict whether α is the encryption of (j, κ_0) or (j, κ_1) — when E asks $\Pi_{j,i}^t$ its **Test** query, A_2 replies with the key κ_0. If E predicts κ_0 is the key held by $\Pi_{j,i}^t$, then A_2 outputs 0 (saying that α is an encryption of $m_0 = (j, \kappa_0)$), otherwise A_2 outputs 1 (saying that α is an encryption of $m_1 = (j, \kappa_1)$).

Further in this event (i.e., where E has picked responder $\Pi_{j,i}^t$ which has had a matching conversation with some $\Pi_{i,j}^s$ to ask its **Test** query), the probability that A_2 has had to decrypt α to answer a **Reveal** query on behalf on some initiator $\Pi_{i,j}^u$ is negligible. For the string that made such a $\Pi_{i,j}^u$ accept must look like:

$$[i, R_j', R_i', \alpha]_{sig_{SSK_j}} \quad .$$

Except with negligible probability, this string has been output by some $\Pi_{j,i}^v$ oracle. If $v \neq t$, then A_2 itself performed the encryption involved in forming the flow, so already knows how to decrypt α and win its experiment. If $v = t$ and $u \neq s$, then $\Pi_{i,j}^s$ and $\Pi_{i,j}^u$ have independently picked the same R_i, which happens with negligible probability. If $v = t$ and $u = s$, then $\Pi_{i,j}^s$ is by definition unopened, and so the problem doesn't arise.

If E does not pick $\Pi_{j,i}^t$ to ask its **Test** query, or picks $\Pi_{j,i}^t$ but $\Pi_{j,i}^t$ has not had a matching conversation with some $\Pi_{i,j}^s$, then A_2 gives up and guesses at random whether α is the encryption of m_0 or m_1.

It can be seen that F built in this way wins its experiment with probability $\frac{1}{2} + \nu(k)$ for $\nu(k)$ non-negligible — contradicting the assumed security of the encryption scheme. We conclude that either $n_2(k)$ or $n_4(k)$ must be negligible.

Case 2. Suppose $n_1(k)$ and $Pr[B_k] = n_3(k)$ are non-negligible, i.e. E picks an initiator oracle with non-negligible probability and in this case succeeds with probability exceeding $\frac{1}{2}$ by a non-negligible amount. We construct from E an adversary $\overline{E}$ of P which with non-negligible probability picks a responder oracle, and then succeeds with probability exceeding $\frac{1}{2}$ by a non-negligible amount.

E's experiment begins — coins are tossed and $\mathcal{G}$ run. $\overline{E}$ is started on input 1^k and *public-info*. $\overline{E}$ now starts E on input 1^k and *public-info*. When E asks one of its oracles a query (other than the **Test** query), $\overline{E}$ simply passes this query on to its own oracles, and relays the reply to E.

$\overline{E}$ answers E's **Test** query as follows. Suppose E picks an initiator oracle $\Pi_{i,j}^s$ to ask its **Test** query. If $\Pi_{i,j}^s$ has not had a matching conversation with some responder $\Pi_{j,i}^t$, then $\overline{E}$ gives up ($\overline{E}$ picks $\Pi_{i,j}^s$ to ask its own **Test** query, and copies E's bit **Guess**) — however this event happens with negligible probability, since P is a secure (mutual) authentication protocol. Otherwise $\Pi_{i,j}^s$ has had a matching conversation with some $\Pi_{j,i}^t$. If $\Pi_{j,i}^t$ has had a matching conversation with both $\Pi_{i,j}^s$ and some other $\Pi_{i,j}^u$ (which is opened), then $\overline{E}$ gives up as before — however this also happens with negligible probability, since in this case $\Pi_{i,j}^s$ and $\Pi_{i,j}^u$ must have both picked the same random value R_i independently. Finally, if $\Pi_{j,i}^t$ has had a matching conversation with only $\Pi_{i,j}^s$, then instead of forwarding E's **Test** query to its own $\Pi_{i,j}^s$, $\overline{E}$ asks its **Test** query to $\Pi_{j,i}^t$ (it is easily seen to be without loss of generality to assume that $\Pi_{j,i}^t$ has accepted). $\overline{E}$ sends $\Pi_{j,i}^t$'s reply to E in place of $\Pi_{i,j}^s$'s response to E's **Test** query, and then uses E's bit **Guess** as its own bit **Guess**.

If E does not pick an initiator oracle to ask its **Test** query, then $\overline{E}$ gives up — $\overline{E}$ picks some initiator oracle to ask its own **Test** query and makes a random selection for **Guess**.

It can be seen that $\overline{E}$ built in this way picks a responder oracle to ask its **Test** query with non-negligible probability, and also succeeds having picked a responder oracle with probability $\frac{1}{2} + \nu(k)$ for non-negligible $\nu(k)$ — this contradicts Case 1 above. We conclude that either $n_1(k)$ or $n_3(k)$ must be negligible.

Together the two cases contradict the assumption that $advantage^E(k)$ is non-negligible. $\qquad\qquad\square$

SG Logic - A Formal Analysis Technique for Authentication Protocols

Sigrid Gürgens *

GMD - German National Research Center for Information Technology
Rheinstrasse 75, D-64295 Darmstadt, Germany
Tel.: +49 6151 869239, Fax: +49 6151 869224
Email: guergens@darmstadt.gmd.de

Abstract. The security of electronic communication relies to a great extent on the security of authentication protocols used to distribute cryptographic keys. Hence formal techniques are needed which help to analyse the security of these protocols. In this paper we introduce a formal method which allows to detect the possibility of certain replay and interleaving attacks. By using our method we are able to show the weakness of the Neuman-Stubblebine protocol and to detect inaccuracies in some authentication protocols standardized in ISO. These inaccuracies may cause the protocol to allow interleaving attacks in certain environments, a fact which seems to be unrecognized so far.

1 Introduction

Confidentiality and authenticity of electronic communication rely to a great extent on the security of authentication protocols used to distribute cryptographic keys. Unfortunately, the literature provides many examples of protocols which were supposed to be correct but later were found to contain flaws (e.g. the Needham-Schroeder protocol [NeSch78], the weakness of which was first shown by Denning and Sacco in [DeSa81], an early version of the CCITT X.509 three way authentication protocol ([X.509-87]), etc.). In particular, there is a number of authentication protocols susceptible to so-called replay and interleaving attacks, where the intruder uses a ciphertext generated by one of the participants taking part in the protocol.

Thus formal analysis techniques are needed, one of the most important ones being the Logic of Authentication by Burrows, Abadi and Needham (the "BAN logic", see [BAN89]), which by reasoning about the beliefs of the principals involved allows to prove certain aspects of protocol security. BAN logic had a great impact on the developement of formal techniques. Quite a number of logics have been published since then. Some of them extend BAN logic to cover different aspects of authentication protocols, such as certificates ([GaSn91]) or key agreement protocols ([vO93]), some improve BAN logic by minor modifications (see,

---

* Funded by Deutsche Telekom AG, Darmstadt

for example, [GNY90]), others take a different approach which is in particular suitable to define a semantic model (e.g. [AbTu91]). However, none of these logics is able to recognize the possibility of a replay or interleaving attack of the type described in this paper.

We introduce a formal method which allows in particular to detect these attacks. This method, which will be called SG logic henceforth, is the revised and extended version of the formalism introduced in [Gue96]. By using the notion of the type of a message we define the message property *not_said*, which enables us to detect the weakness of the Neuman-Stubblebine protocol shown by Syverson in [Syv93] and independently by Carlsen in [Ca93]. Furthermore, we show that the three pass authentication protocol 5.2.2 standardized in ISO/IEC 9798-2 ([9798-2]) contains inaccuracies which may cause the protocol to allow interleaving attacks in certain environments, a fact which seems not to be detectable using other BAN-like logics.

In the next section the Neuman-Stubblebine protocol and the attack is given, thus emphasizing the importance of the questions discussed in this paper. Section 3 describes the semantical model as far as it is necessary to understand the syntax, section 4 explains the syntax and the inference rules. The analysis of the Neuman-Stubblebine protocol is given in section 5, that of the ISO authentication protocol in section 6. Finally, in section 7 we give our conclusions.

2 An example

In what follows we assume that the reader is familiar with the usually used notation for cryptographic protocols and with the mechanisms used to ensure privacy and authenticity. Let us consider a protocol introduced by Neuman and Stubblebine in [NeSt93] and a possible attack:

1. $A \longrightarrow B : A, N_A$
2. $B \longrightarrow S : B, \{K_{BS}; A, N_A, T_B\}, N_B$
3. $S \longrightarrow A : \{K_{AS}; B, N_A, K_{AB}, T_B\}, \{K_{BS}; A, K_{AB}, T_B\}, N_B$
4. $A \longrightarrow B : \{K_{BS}; A, K_{AB}, T_B\}, \{K_{AB}; N_B\}$

The crucial point of this protocol is step 4. After receiving the last message B deciphers the first ciphertext and checks that the first datafield of the result is equal to A and that the third datafield is equal to T_B. He then takes the second datafield to be the new session key, uses it to decipher the second ciphertext and checks that the result is equal to N_B. Finally he believes that he owns a session key newly generated by S which he shares with A.

The attack described e.g. in [Syv93] shows that the above protocol may fail to achieve its aim. In this attack the attacker E impersonates A in the first message and B replies according to the protocol description as he has no means to control who actually sent the message:

1. $E_A \longrightarrow B : A, N_E$
2. $B \longrightarrow E_S : B, \{K_{BS}; A, N_E, T_B\}, N_B$

E intercepts $B's$ message to S, thus the third message is omitted, and sends B the following fourth message, pretending again to be A:

4. $E_A \longrightarrow B : \{K_{BS}; A, N_E, T_B\}, \{N_E; N_B\}$

B checks that the first part of the ciphertext equals A and that the last part equals T_B, which they do. Thus B takes N_E to be the new session key and falsely believes he shares this key with A.

The attack succeeds because of two properties of the above protocol: First, B can not distinguish between (A, N_E, T_B) and (A, K_{AB}, T_B), since he only checks A's identifier and T_B. Second, when using a symmetric key either of the principals in possession of this key can have generated the ciphertext. In other words, when receiving the ciphertext generated with K_{BS}, there is no way for B to tell whether it was generated by S or by himself.

The above attack can be avoided by simple modifications of the protocol, such as using direction bits, reversing the order of N_A, T_B in message two, or requiring B to check that $N_A \neq K_{AB}$ etc. (for a detailed discussion of this see [Syv93]). However, these measures usually add to the cost of performing the protocol (more data has to be encrypted, more checks have to be performed) and in certain environments it might be desirable to restrict to those measures absolutely necessary to provide security. Furthermore we can not be sure that a protocol using these techniques will not be susceptible to this type of attack. In section 6 we show the weakness of a protocol that can not be avoided using typing.

3   The semantic model

The underlying semantic model is only described as far as it is necessary to understand the syntax.

A cryptographic protocol consists of finitely many steps 1 to n. In every step a message is sent by one principal and received by another principal, where the receiving principal does or does not perform some checks on the message. We denote the performance of all steps of a protocol with *protocol run*.

3.1   Notations

Message: The smallest possible message can be a nonce, a timestamp, a principal's identifier, a key, a text field, i.e. a message which can not be further dissected. We call those messages *atomic messages*. Furthermore, a ciphertext is a message (but not an atomic message). Messages can be concatenated to construct new messages: If m and n are messages, so is the concatenation (m, n). However, (m, n) is not equal to (n, m).

In what follows we will use the notion of *recognizable messages*. A message m is recognizable by principal P if m is stored in P's local environment. Moreover we use the following notations:

$\mathcal{P}$: denotes the set of principals and principals' identifiers.

$[P, \ldots, Q]$: denotes a subset of $\mathcal{P}$. (We use square brackets instead of the usually used braces to distinguish clearly a ciphertext from a set of principals.)

$\{K; m\}$: denotes the message m, enciphered or signed with the key K.

$h(m)$: denotes the hash value of the message m.

K^{-1} : denotes the key inverse to K.

K_{AB} : is the usual way to denote a symmetric key shared by the principals A and B. Observe, however, that no logical statements can be concluded as to who actually owns this key.

(r, s) : denotes step s of run r of the given protocol. $s \leq s'$ iff $s = s'$ or step s is performed before step s' and $r \leq r'$ iff $r = r'$ or the first step of r is performed before the first step of r'. Note that protocol run r' may possibly end before run r ends.

$\mathcal{T}(P, (r, s))$: denotes the set of all time variant parameters such as nonces or timestamps principal P accepts at step s of protocol run r.

$\mathcal{R}(P, (r, s))$: denotes the set of all messages recognizable by P at step s of protocol run r.

$\mathcal{G}(P, (r, s))$: denotes the set of messages possibly generated by P before step s of protocol run r is performed.

$\mathcal{G}(\mathcal{P}, (r, s))$: denotes the set of all messages possibly generated by any of the principals before step s of protocol run r is performed.

Unlike BAN and its modifications which divide time into "before the current protocol run" and "during the current protocol run", the set $\mathcal{T}(P, (r, s))$ allows to describe precisely the way time variant parameters like nonces and timestamps are used in authentication protocols: They ensure the receiving principal not only that the message was generated during the current protocol run but also that it was generated until a certain step of this protocol run. $\mathcal{T}(P, (r, s))$ is determined as follows. If principal P generates and sends a nonce N_P in step k of protocol run r and accepts it in a later step $k' > k$, then $N_P \in \mathcal{T}(P, (r, s))$ for $k \leq s \leq k'$. If he accepts a timestamp T_Q from step k to step k', then $T_Q \in \mathcal{T}(P, (r, s))$ for $k \leq s \leq k'$.

The set $\mathcal{R}(P, (r, s))$ is determined as follows. If principal P receives messages m in (r, s) and m' in (r, s'), where $s \leq s'$, and the protocol description requires him to check that $m = m'$ or $m \neq m'$, then $m \in \mathcal{R}(P, (r, s'))$. If P receives a message m in (r, s) and the protocol description requires him to check whether $m \in \mathcal{P}$ or $m \in \mathcal{T}(P, (r, s))$ then $m \in \mathcal{R}(P, (r, s))$.

A protocol description does not only consist of the messages to be sent but also of the description of the checks to be performed by the recipients of the messages. The following definition of the type of a message captures those checks.

Definition 1 *Let $P \in \mathcal{P}$ be a principal, m be an atomic message. The type τ of message m at step s of protocol run r with respect to principal P is defined as follows:*

$$\tau(P, m, (r,s)) := \begin{cases} n\bar{n}_1 \ldots \bar{n}_r & \text{if } n, n_1, \ldots, n_r \in \mathcal{R}(P, (r,s)) \text{ and } P \\ & \text{verifies that } m = n \text{ or} \\ & m \neq n_j \text{ for all } j \in \{1, \ldots, r\} \\ \\ \Box & \text{otherwise} \end{cases}$$

For a message $m = (m_1, \ldots, m_n)$ the type is given as

$$\tau(P, (m_1, \ldots, m_n), (r,s)) := \tau(P, m_1, (r,s)), \ldots, \tau(P, m_n, (r,s))$$

The type of a hash value is defined as:

$$\tau(P, h(m_1, \ldots, m_n), (r,s)) := h(\tau(P, m_1, (r,s)), \ldots, \tau(P, m_n, (r,s)))$$

and the type of a ciphertext is defined as:

$$\tau(P, \{K; m_1, \ldots, m_n\}, (r,s)) := \begin{cases} (K, \tau(P, m_1, (r,s)), \ldots, \tau(P, m_n, (r,s))) \\ \text{if } P \text{ owns } K^{-1} \text{ and applies it to} \\ \{K; m_1, \ldots, m_n\} \\ \\ \Box \text{ otherwise} \end{cases}$$

Two messages have the same type with respect to principal P if P can not distinguish between them.

$\mathcal{G}(\mathcal{P}, (r,s))$ is the set of messages principal P has generated or might have generated before step s of protocol run r. Hence $\mathcal{G}(P, (r,1)) \subseteq \mathcal{G}(P, (r,s))$ includes all messages P generated in previous protocol runs. Furthermore, every message P generated from steps 1 to $s - 1$ is element of $\mathcal{G}(P, (r,s))$. In addition to that, P may have generated further messages. Let us assume that meanwhile a second protocol run r' was started and P expects the message m in step s' of this run. Let us further assume that there exists a message $m' \in \mathcal{G}(\mathcal{P}, (r,s))$ which has exactly the same type as message m (i.e. $\tau(P, m', (r', s')) = \tau(P, m, (r', s'))$). Then P accepts message m' in step s' of protocol run r' and responds according to the protocol description, i.e. generates and sends message m'' in the next step $s'' > s'$ of protocol run r'. Since at this point protocol run r has not yet terminated (the next step to be performed is s), m'' is a message that P may have generated before point (r,s) and $m'' \in \mathcal{G}(P, (r,s))$. Obviously, if $(m,n) \in \mathcal{G}(P, (r,s))$, then $m \in \mathcal{G}(P, (r,s))$ and $n \in \mathcal{G}(P, (r,s))$.

The set $\mathcal{G}(P, (r,s))$ in particular captures those messages principal P might be caused to generate in an interleaving attack. In this type of attack the intruder typically uses a message of the protocol run r_1 to start an interleaving protocol run r_2, where he uses the principal P to be attacked as an oracle. The ciphertext hereby generated is then sent to P in the protocol run r_1.

4 The logic

4.1 The syntax

In the following, **P** denotes the set of principals and principals' names, respectively, **K** the set of keys and $\mathbb{P}(\mathbf{P})$ the power set of **P**.

The functions and predicates of SG logic have the following semantic meaning:

$\varepsilon : \mathbf{K} \longrightarrow \mathbb{P}(\mathbf{P})$: The function ε maps a key K onto the set of all principals who own K or will own K after the protocol run and can use it to check the authenticity of a ciphertext generated with K^{-1} (i.e. who can use it to check digital signatures and message authentication codes, respectively).

$\delta : \mathbf{K} \longrightarrow \mathbb{P}(\mathbf{P})$: The function δ maps a key K onto the set of principals who own K or will own K after the protocol run and can use it to generate signatures and message authentication codes, respectively, and to decipher messages enciphered with K^{-1}.

$in(P, V)$: This predicate is used to express the fact that the principal P is element of $\varepsilon(K)$ or $\delta(K)$, i.e. that P can use the key K according to its purpose.

$number(V, W)$: holds if $V, W \in \mathbb{P}(\mathbf{P})$ and the sets V and W have the same cardinality.

$leq(x, y)$: holds if x and y are protocol steps and $x = y$ or step y is performed after step x.

$enc(K)$: holds if K is a key for both encryption and checking digital signatures and message authentication codes, respectively, generated with K^{-1}.

$sig(K)$: holds if K is a key for generating digital signatures and message authentication codes, respectively, and for decrypting messages encrypted with K^{-1}.

$sees(P, m, r, s)$: holds if P receives message m in step s of the protocol run r.

$has(P, m)$: holds if P has at some point received the message m and owns it.

$said(P, K, m, r, s)$: holds if P has sent, but not received message $\{K; m\}$ until or at step s of protocol run r.

$not\_said(P, K, m, r, s)$: holds if there is no message m' in the set of possibly by P until (r, s) generated messages having the same type as $\{K; m\}$ with respect to P and (r, s), i.e. if $\forall m' \in \mathcal{G}(P, (r, s))$: $\tau(P, m', (r, s)) \neq \tau(P, \{K; m\}, (r, s))$.

$recog(P, m, r, s)$: holds if P recognizes message m in (r, s), i.e. if $m \in \mathcal{R}(P, (r, s))$.

$in\_time(P, m, r, s)$: holds if P accepts the message m at (r, s) as having a certain timeliness property, i.e. if $m \in \mathcal{T}(P, (r, s))$.

$determ(P, m)$: holds if the message m is determined by P, which means in particular that P did not receive it in some earlier step.

$says(P, K, m)$: holds if P said m using the key K during an accepted time interval (e.g. during the current protocol run).

$controls(P, \varphi)$: holds if P can be trusted as to statements φ.

$believes(P, \varphi)$: holds if P believes φ and acts accordingly.

The constructs *believes* and *controls* are adopted from BAN logic, *sees* and *said* are BAN constructs and *says* is a construct adopted from Abadi and Tuttle, modified with respect to (r, s) if necessary.

With the functions E and D, which are the semantic equivalent to ε and δ, a key can be represented as an element of $\mathbb{P}(\mathcal{P}) \times \mathbb{P}(\mathcal{P})$. This enables us to describe keys for both symmetric and asymmetric algorithms, and, by means of inference rules, to take into account the peculiarities of the respective algorithm. A symmetric key K_{PQ} provides authenticity as well as privacy and additionally $K_{PQ} = K_{PQ}^{-1}$, hence we have $E(K_{PQ}) = D(K_{PQ}) = E(K_{PQ}^{-1}) = D(K_{PQ}^{-1}) = [P, Q]$. Since the public key algorithm RSA has the property that a ciphertext generated with a private key can be deciphered using the corresponding public key ($\{K_P; \{K_P^{-1}; m\}\} = m$) (see [RSA78]), a public RSA key K_P of P is described as $E(K_P) = D(K_P) = \mathcal{P}$, P's private key K_P^{-1} as $E(K_P^{-1}) = D(K_P^{-1}) = [P]$. In contrast to this we have $D(K_P) = \emptyset$ and $E(K_P^{-1}) = \emptyset$ if (K_P, K_P^{-1}) is a key pair for an algorithm where the application of the private key is a one way function (see e.g. [ElGa85]).

4.2 Inference Rules

The following inference rules do not use universal quantification, so $P, Q, \ldots$ should be read as $\forall P, \forall Q, \ldots$, K, K^{-1} as $\forall K, \forall K^{-1}$, the respective statements hold for the notation of protocol steps $(s, s', \ldots)$ and runs $(r, r', \ldots)$ and statements $(\varphi_1, \varphi_2, \ldots)$.

The rules describe the mechanisms used in authentication protocols and the conclusions that can be drawn using these mechanisms. Observe that these rules can be applied to both symmetric and asymmetric protocols (that is, protocols that use symmetric and asymmetric algorithms, respectively). They are based on the general assumptions that all cryptographic algorithms are secure, that the ciphertext $\{K; m\}$ can only be decrypted using K^{-1}, that a cryptographic key can not be guessed etc.

- The **plaintext recovery rule** (PR) describes when principal P can decipher a ciphertext. For this he must have received the ciphertext (i.e. he must "see" it) and he must own the key needed for deciphering.
  The fact that for deciphering no belief about the quality of the key is needed was first mentioned in [GNY90].

$$(PR) \quad \frac{sees(P, \{K; m\}, r, s) \ \wedge \ in(P, \delta(K^{-1}))}{sees(P, m, r, s)}$$

- The **message meaning rules** (MM1) and (MM2) together with rule (S) describe the way the recipient of a ciphertext can draw conclusions about its generator. P must have received the ciphertext and must own the respective cleartext as well. Trivially, when using a symmetric algorithm this assumption holds if P owns the key necessary for checking the authenticity of the ciphertext. However, when using an asymmetric algorithm where the digital signature can not be retransformed into the cleartext this assumption has to be checked explicitly.
  Furthermore, part of the message has to be recognizable by P. (This corresponds to the assumption P *believes* $\phi(X)$ in [GNY90].) Of course P needs to own the key necessary for the authenticity check ($in(P, \varepsilon(K))$) and has to have an idea about who owns the key used for generating the ciphertext ($believes(P, \delta(K^{-1}) = [P_1, \ldots, P_n])$). Then P can draw the conclusion that each of the latter can have generated the ciphertext (rule (MM1)).

$$(MM1) \quad \frac{\begin{array}{c} sees(P, \{K^{-1}; m_1, \ldots, m_n\}, r, s) \ \wedge \ has(P, (m_1, \ldots, m_n)) \\ \wedge \ (recog(P, m_1, r, s) \ \vee \ \ldots \ \vee \ recog(P, m_n, r, s)) \\ \wedge \ in(P, \varepsilon(K)) \ \wedge \ believes(P, \delta(K^{-1}) = [P_1, \ldots, P_n]) \end{array}}{\begin{array}{c} believes(P, said(P_1, K^{-1}, (m_1, \ldots, m_n), r, s) \ \vee \ \ldots \\ \ldots \ \vee \ said(P_n, K^{-1}, (m_1, \ldots, m_n), r, s)) \end{array}}$$

If analysing an asymmetric protocol rule (MM1) is sufficient to conclude that the owner of the private key generated the digital signature. However, when using a symmetric key K, in general we have $believes(P, \delta(K) = [P, Q])$ and (MM1) merely allows the conclusion that either P or Q can have generated the ciphertext. In this case we need an additional inference rule which allows to conclude when P can not have generated the ciphertext. This is true if every ciphertext in the set of messages possibly generated by P has a type different to the type of the ciphertext in question.

$$(\text{MM2}) \quad \frac{not\_said(P, K, m, r, s)}{believes(P, \neg(said(P, K, m, r, s)))}$$

- The idea of our **nonce verification rule** (NV) corresponds to that of [BAN89] and [AbTu91], respectively: If P believes that the message m was generated by Q and has certain timeliness properties then he can believe that it was generated during an accepted time interval. This can either be the current protocol run or, if dealing with a certificate, the time interval from generating the certificate to the current moment. Once the message is accepted as being generated during a certain time interval the actual protocol step and run of the generation are no longer of interest.

$$(\text{NV}) \quad \frac{\begin{array}{c} believes(P, said(Q, K, (m_1, \ldots, m_n), r, s)) \\ \wedge \; (in\_time(P, m_1, r, s) \; \vee \; \ldots \; \vee \; in\_time(P, m_n, r, s) \\ \vee \; in\_time(P, K, r, s)) \end{array}}{believes(P, says(Q, K, (m_1, \ldots, m_n)))}$$

- In certain cases we need the following rule for dealing with hash values:

$$(\text{HS}) \quad \frac{\begin{array}{c} believes(P, says(Q, K, h(m))) \; \wedge \; has(P, m) \\ \wedge \; determ(Q, h(m)) \end{array}}{believes(P, says(Q, K, m))}$$

When using this inference rule it has to be considered carefully whether or not the predicate $determ(Q, h(m))$ holds. For this, one should take the possibility into account that Q might have received $h(m)$ "hidden", as part of a different message without realizing this.

- The **jurisdiction rule** (J1) is basically the same as in BAN and its modifications and describes the special role of a key server: If P believes that Q (the key server) is trustworthy with respect to the statement φ (e.g. a statement about a session key), and if P believes that this statement was generated within an accepted time interval (e.g. during the current protocol run), then P can believe the statement φ.

$$(\text{J1}) \quad \frac{believes(P, says(Q, K, \varphi)) \; \wedge \; believes(P, controls(Q, \varphi))}{believes(P, \varphi)}$$

The jurisdiction rule (J2) covers those cases where the principal P both generates and uses the key to be exchanged in the protocol (i.e. those cases where P is not a key server).

$$(\text{J2}) \quad \frac{\begin{array}{c} believes(P, says(Q, K', K)) \\ \wedge \; believes(P, controls(Q, in(Q, \delta(K^{-1})))) \; \wedge \; sig(K) \end{array}}{believes(P, in(Q, \delta(K^{-1})))}$$

- The following rules specify when a principal owns a certain message or a certain key. The latter is needed for protocols where P uses the session key just exchanged to indicate its possession.

$$(\text{H1}) \quad \frac{believes(P, says(Q, K, m))}{believes(P, has(Q, K))} \qquad (\text{H3}) \quad \frac{has(P, m)}{has(P, h(m))}$$

$$(\text{H2}) \quad \frac{believes(P, says(Q, K, m))}{believes(P, has(Q, m))} \qquad (\text{H4}) \quad \frac{sees(P, m, r, s)}{has(P, m)}$$

- The **key rules** (K1) and (K2) reflect the fact that a principal can use any key he owns. Of course, a key can only be used according to its purpose (e.g. a private key for the ElGamal signature algorithm can not be used to check the authenticity of a digital signature).

  Inference rule (K3) describes how to interpret the name R of a principal connected with a key K as part of a ciphertext: The generator of this message wants to express the fact that R owns (or will own) the key K^{-1} corresponding to K. Rule (K4) allows to draw the conclusion about the set of principals that can generate a message authentication code and a digital signature, respectively, using K.

$$(\text{K1}) \quad \frac{has(P, K) \ \wedge \ enc(K)}{in(P, \varepsilon(K))}$$

$$(\text{K2}) \quad \frac{has(P, K) \ \wedge \ sig(K)}{in(P, \delta(K)) \ \wedge \ believes(P, in(P, \delta(K)))}$$

$$(\text{K3}) \quad \frac{\begin{array}{c}(believes(P, says(Q, K', (R, K))) \\ \vee \ believes(P, says(Q, K', (K, R)))) \ \wedge \ sig(K^{-1})\end{array}}{believes(P, says(Q, K', in(R, \delta(K^{-1}))))}$$

$$(\text{K4}) \quad \frac{\begin{array}{c}believes(P, in(P_1, \delta(K)) \ \wedge \ \ldots \ \wedge \ in(P_n, \delta(K))) \\ \wedge \ number([P_1, \ldots, P_n], \delta(K))\end{array}}{believes(P, \delta(K) = [P_1, \ldots, P_n])}$$

- **Recognizability rules**: Under certain assumptions the recognizability of a message transfers:

$$(\text{R1}) \quad \frac{recog(P, m, r, s)}{recog(P, h(m), r, s)}$$

$$(\text{R2}) \quad \frac{recog(P, m, r, s) \ \wedge \ in(P, \varepsilon(K))}{recog(P, \{K; m\}, r, s) \ \wedge \ recog(P, \{K^{-1}; m\}, r, s)}$$

$$(\text{R3}) \quad \frac{recog(P,m,r,s) \;\wedge\; in(P,\delta(K))}{recog(P,\{K;m\},r,s)}$$

$$(\text{R4}) \quad \frac{recog(P,m_1,r,s) \;\wedge\; recog(P,m_2,r,s)}{recog(P,(m_1,m_2),r,s)}$$

- **Time rules**: Under certain assumptions the timeliness of a message transfers:

$$(\text{T1}) \quad \frac{in\_time(P,m,r,s)}{in\_time(P,h(n_1,\ldots,m,\ldots,n_k),r,s)}$$

$$(\text{T2}) \quad \frac{\begin{array}{c} in\_time(P,m_i,r,s) \;\wedge\; in(P,\varepsilon(K^{-1})) \\ \wedge\; \neg(m_i = \{K^{-1};m_1,\ldots,m_k\}) \end{array}}{in\_time(P,\{K;m_1,\ldots,m_k\},r,s)}$$

$$(\text{T3}) \quad \frac{in\_time(P,K,r,s)}{in\_time(P,\{K;m\},r,s)}$$

Rule (T2) takes an observation into account that was mentioned in [MSN94]: If a ciphertext is accepted as to its timeliness because the key used has this property, but the enciphered message does not have it, then it is not allowed to conclude that $m = \{K; \{K^{-1};m\}\}$ can be accepted with respect to timeliness.

- For technical reasons we need the following **composition** and **decomposition rules**:

$$(\text{C}) \quad \frac{has(P,m_1) \;\wedge\; has(P,m_2)}{has(P,(m_1,m_2))}$$

$$(\text{D1}) \quad \frac{sees(P,(m_1,m_2),r,s)}{sees(P,m_1,r,s) \;\wedge\; sees(P,m_2,r,s)}$$

$$(\text{D2}) \quad \frac{has(P,(m_1,m_2))}{has(P,m_1) \;\wedge\; has(P,m_2)}$$

$$(\text{D3}) \quad \frac{believes(P,said(Q,K,(m_1,m_2),r,s))}{\begin{array}{c} believes(P,said(Q,K,m_1,r,s)) \;\wedge\; \\ believes(P,said(Q,K,m_2,r,s)) \end{array}}$$

$$(\text{D4}) \quad \frac{believes(P,says(Q,K,(m_1,m_2)))}{\begin{array}{c} believes(P,says(Q,K,m_1)) \;\wedge\; \\ believes(P,says(Q,K,m_2)) \end{array}}$$

Observe that neither do $believes(P,said(Q,K,m_1,r,s))$ and $believes(P,said(Q,K,m_2,r,s))$ imply $believes(P,said(Q,K,(m_1,m_2),r,s))$ nor can the equivalent conclusion for *says* be drawn.

- Since we assume that the reader is familiar with the usual rules of first order and modal logic we only mention that rule for statements $\varphi_1, \ldots, \varphi_n$ which is actually needed in the following proofs.

$$\text{(S)} \quad \frac{\begin{array}{c} believes(P, \varphi_1 \vee \ldots \vee \varphi_n) \wedge \\ believes(P, \neg\varphi_1 \wedge \ldots \wedge \neg\varphi_{i-1} \wedge \neg\varphi_{i+1} \wedge \ldots \wedge \neg\varphi_n) \end{array}}{believes(P, \varphi_i)}$$

5 The analysis

We are now ready to analyse the Neuman-Stubblebine protocol given in section 2. We will restrict ourselves to analyse the security of B, thus proving formally that the attack explained in section 2 is possible. In other words, we want to show, using SG logic, that the ciphertext B accepts in step 4 of the protocol is not necessarily generated by S. This is done by showing that there is a similar ciphertext in the set of messages possibly generated by B having the same type.

Before starting the actual analysis, though, it is necessary to determine the sets $\mathcal{T}(B, (r,s))$, $\mathcal{R}(B, (r,s))$ and $\mathcal{G}(B, (r,s))$ as well as the assumptions which hold before the start of the current protocol run r.

Since B receives message 4 we have

$$sees(B, (\{K_{BS}; A, K_{AB}, T_B\}, \{K_{AB}; N_B\}), r, 4) \tag{1}$$

B sends message $(B, \{K_{BS}; A, N_A, T_B\}, N_B)$ in step 2 of the protocol run, so this message is element of the set of messages possibly sent by B. In particular,

$$\{K_{BS}; A, N_A, T_B\} \in \mathcal{G}(B, (r, 4)) \tag{2}$$

The keys used in the protocol are symmetric keys, so we have

$$K_{BS} = K_{BS}^{-1} \tag{3}$$
$$K_{AB} = K_{AB}^{-1} \tag{4}$$
$$sig(K_{AB}) \tag{5}$$
$$number([P, Q], \delta(K_{AB})) \tag{6}$$

(K_{AB} being a symmetric key it follows that it is a key for checking authenticity and that it will be owned by only two principals after the protocol run.)

K_{BS} is a symmetric authentication key shared by B and S, yielding

$$in(B, \delta(K_{BS}^{-1})) \tag{7}$$
$$in(B, \varepsilon(K_{BS}^{-1})) \tag{8}$$
$$believes(B, \delta(K_{BS}^{-1}) = [B, S]) \tag{9}$$

The key server is trusted as to statements about session keys, so

$$believes(B, controls(S, in(P, \delta(K)))) \tag{10}$$

Since B sends the timestamp T_B and the nonce N_B in step 2 and accepts them in step 4 we have $\mathcal{T}(B, (r, 4)) = \{T_B, N_B\}$. It follows in particular that

$$in\_time(P, T_B, r, 4) \tag{11}$$

The protocol description in [NeSt93] requires B to check A, T_B and N_B at step 4 of the protocol run. However, it does not demand of B to perform any checks on K_{AB} at step 4, in particular it does not require B to check that $N_A \neq K_{AB}$. Hence $N_A \notin \mathcal{R}(B, (r, 4))$ and $\mathcal{R}(B, (r, 4)) = \{A, T_B, N_B\}$. This yields

$$recog(B, A, r, 4) \tag{12}$$

After these preliminaries we can start the actual analysis. One of the goals of the protocol is that after its run B shares the session key K_{AB} with A. Thus, we have to prove $believes(B, \delta(K_{AB}) = [A, B])$.

$$(1) \quad \xrightarrow{D1} \quad sees(B, \{K_{BS}; A, K_{AB}, T_B\}, r, 4) \tag{13}$$

$$(13),(7) \quad \xrightarrow{PR} \quad sees(B, (A, K_{AB}, T_B), r, 4) \tag{14}$$

$$(14) \quad \xrightarrow{H4} \quad has(B, (A, K_{AB}, T_B)) \tag{15}$$

$$\begin{aligned}(3),(8),(9), \quad &\xrightarrow{MM1} \quad believes(B, said(B, K_{BS}^{-1}, (A, K_{AB}, T_B), r, 4) \\ (12),(13),(15) \quad & \qquad \vee \; said(S, K_{BS}^{-1}, (A, K_{AB}, T_B), r, 4)) \end{aligned} \tag{16}$$

For being able to conclude that it was indeed S who said (A, K_{AB}, T_B) at $(r, 4)$ we need to find out whether or not $not\_said(B, K_{BS}^{-1}, (A, K_{AB}, T_B), r, 4)$ holds, i.e. whether or not there exists a message $m \in \mathcal{G}(B, (r, 4))$ with $\tau(B, m, (r, 4)) = \tau(B, \{K_{BS}; A, K_{AB}, T_B\}, r, 4)$.

As we already pointed out, $\{K_{BS}; A, N_A, T_B\} \in \mathcal{G}(B, (r, 4))$ (see (2)). Furthermore, $\tau(B, \{K_{BS}; A, N_A, T_B\}, (r, 4)) = (K_{BS}, A, \square, T_B) = \tau(B, \{K_{BS}; A, K_{AB}, T_B\}, r, 4)$. So there is indeed a message in the set of possibly generated messages with the same type as the type of the ciphertext in question, in other words, $not\_said(B, K_{BS}^{-1}, (A, K_{AB}, T_B), r, 4)$ does not hold.

Hence we are not able to apply rule (MM2) and therefore can not continue the analysis to prove the protocol secure for B. The fact that $\{K_{BS}; A, N_A, T_B\} \in \mathcal{G}(B, (r, 4))$ has the same type as the ciphertext which B receives in step 4 of the protocol run shows that B is not able to distinguish between the two. Thus an intruder can replay the ciphertext generated by B for the second protocol step in the fourth message without B noticing, showing that the attack described in section 2 is possible.

Of course we are not only interested in finding the weakness of a protocol but also in finding a way to repair it. In the above case this can be easily achieved. The protocol has to be changed in a way that the two ciphertexts in question do not have the same type (with respect to B). This can be attained by requiring B to check $N_A \neq K_{AB}$ after having received the fourth message. A second possibility was suggested in [Syv93]: We just need to switch the order of N_A, T_B in the second message. This solution has the advantage that it does not need an additional check to be performed by B.

In this case, the second message of the protocol becomes $(B, \{K_{BS}; A, T_B,$

$N_A\}, N_B)$ which yields $\tau(B, \{K_{BS}; A, T_B, N_A\}, (r,4)) = (K_{BS}, A, \square, \square) \neq (K_{BS}, A, \square, T_B) = \tau(B, \{K_{BS}; A, K_{AB}, T_B\}, (r,4))$. Since there is no other ciphertext in the protocol with three components there is no message in the set of possibly by B generated messages having the same type as $\{K_{BS}; A, K_{AB}, T_B\}$, i.e.

$$not\_said(B, K_{BS}^{-1}, (A, K_{AB}, T_B), r, 4) \tag{17}$$

holds and we can continue the analysis.

$$(17) \xrightarrow{MM2} believes(B, \neg(said(B, K_{BS}^{-1}, (A, K_{AB}, T_B), r, 4))) \tag{18}$$
$$(16),(18) \xrightarrow{S} believes(B, said(S, K_{BS}^{-1}, (A, K_{AB}, T_B), r, 4)) \tag{19}$$
$$(11),(19) \xrightarrow{NV} believes(B, says(S, K_{BS}^{-1}, (A, K_{AB}, T_B))) \tag{20}$$
$$(20) \xrightarrow{D4} believes(B, says(S, K_{BS}^{-1}, (A, K_{AB}))) \tag{21}$$
$$(4),(5),(21) \xrightarrow{K3} believes(B, says(S, K_{BS}^{-1}, in(A, \delta(K_{AB})))) \tag{22}$$

$$(10),(22) \xrightarrow{J1} believes(B, in(A, \delta(K_{AB}))) \tag{23}$$
$$(14) \xrightarrow{D1} sees(B, K_{AB}, r, 4) \tag{24}$$
$$(24) \xrightarrow{H4} has(B, K_{AB}) \tag{25}$$
$$(5),(25) \xrightarrow{K2} believes(B, in(B, \delta(K_{AB}))) \tag{26}$$
$$(6),(23),(26) \xrightarrow{K4} \boxed{believes(B, \delta(K_{AB}) = [A, B])} \tag{27}$$

We stop the analysis at this point as the above is sufficient to show how our method works. In particular it shows that the new concept of the type of a message is suitable to describe the checks which are performed by the principal receiving a message. Two messages having the same type are not distinguishable, so any protocol using such messages is vulnerable to a replay attack as described in section 2.

However, one should be aware that the fact that the above analysis does not find any other weakness in the modified protocol does not necessarily mean that it is secure. Unfortunately the logic is only able to show the weakness of a protocol but not its security.

In the next section we will show that the type of a message can be used equally well to describe which messages are accepted by a principal, hereby proving the possibility of an interleaving attack in a protocol standardized in ISO.

6 The three pass authentication protocol of ISO 9798-2

Let us now consider the following authentication protocol standardized by ISO in ISO/IEC 9798-2 [9798-2].

1. $B \longrightarrow A : N_B, Text1$
2. $A \longrightarrow B : Text3, \{K_{AB}; N_A, N_B, B, Text2\}$
3. $B \longrightarrow A : Text5, \{K_{AB}; N_B, N_A, Text4\}$

This is a protocol template rather than a precise description of the messages to be sent. [9798-2] states that text fields may be used for various purposes. They may identify the key to be used for encryption, contain a timestamp or any other data or they may be omitted. [9798-2] gives no hint as to how principal A recognizes that he communicates with B nor does it explain the way A determines the key to be used.

[9798-2] states that the distinguishing identifier "B" is included in the ciphertext of the second step to prevent a so-called reflection attack, where an intruder reflects B's nonce N_B to B, pretending to be A. However, we will show that even if including "B", an attack on B is possible in a certain environment.

Let us assume that the protocol is used in an environment where the principals are not able to recognize each other's identity, where the key K_{AB} to be used by A is not determined by the messages of the protocol, and which allows interleaving protocol runs. One instance for such an environment is the communication between a smartcard and a smartcard reader, where in particular the key to be used by the smartcard can be determined in a parameter of the smartcard command (see [7816-4]).

Since smartcards can not determine who sent the first message it is obvious to use the text field $Text1$ to indicate this. Omitting text fields $Text2$ and $Text3$, the protocol becomes:

1. $B \longrightarrow A : N_B, B$
2. $A \longrightarrow B : \{K_{AB}; N_A, N_B, B\}$
3. $B \longrightarrow A : \{K_{AB}; N_B, N_A\}$

Analogous to the analysis in section 5 we can conclude, using the assumptions $sees(B, \{K_{AB}; N_A, N_B, B\}, r_1, 2)$, $K_{AB} = K_{AB}^{-1}$, $in(B, \delta(K_{AB}^{-1}))$, $in(B, \varepsilon(K_{AB}))$, $believes(B, \delta(K_{AB}^{-1}) = [A, B])$ and $recog(B, B, r_1, 2)$, that $believes(B, said(A, K_{AB}^{-1}, (N_A, N_B, B), r_1, 2) \vee said(B, K_{AB}^{-1}, (N_A, N_B, B), r_1, 2))$ holds.

Let us now focus for a moment on the set of messages possibly generated by B before step 2 of protocol run r_1. At this point B accepts two different messages: the second message of protocol run r_1 and the first message of any interleaving protocol run r_2. Since the protocol description in [9798-2] does not require B to perform any checks on the first message of a protocol run, B will accept any principal's distinguishing identifier, even if it is his own, i.e. $\mathcal{R}(B, (r_2, 1)) = \emptyset$ and $\tau(B, m, (r_2, 1)) = \square, \square$ for any message m consisting of two parts received at $(r_2, 1)$.

Since B sends the message (N_B, B) in the first step of protocol run r_1, $(N_B, B) \in \mathcal{G}(B, (r_1, 2)) \subseteq \mathcal{G}(\mathcal{P}, (r_1, 2)) \subseteq \mathcal{G}(\mathcal{P}, (r_2, 1))$ for every protocol run r_2 started interleavingly $(r_2 > r_1)$. So the message (N_B, B) is generally known. Furthermore, $\tau(B, (N_B, B), (r_2, 1)) = \square, \square$. In other words, if an intruder E starts a second protocol run r_2 with sending B the message (N_B, B), B accepts this and replies with the appropriate response $\{K_{AB}; N_B', N_B, B\}$. (Note that by assumption the key K_{AB} is determined by means other than the protocol messages.) So $\{K_{AB}; N_B', N_B, B\}$ is one of the messages possibly generated by B before step 2 of the first protocol run r_1 is performed ($\{K_{AB}; N_B', N_B, B\} \in \mathcal{G}(B, (r_1, 2))$), as run r_1 has not yet terminated.

Again, for being able to proceed, we need to find out whether $not\_said(B, K_{AB}^{-1}, (N_A, N_B, B), r_1, 2)$ holds. B checks the second and the third datafield and $\mathcal{R}(B, (r_1, 2)) = \{N_B, B\}$, which yields $\tau(B, \{K_{AB}; N_A, N_B, B\} = (K_{AB}, \Box, N_B, B)$. As we have already pointed out, we have $\{K_{AB}; N_B', N_B, B\} \in \mathcal{G}(B, (r_1, 2))$, where $\tau(B, \{K_{AB}; N_B', N_B, B\, (r_1, 2)) = (K_{AB}, \Box, N_B, B)$. So again there exists a message in the set of messages possibly generated by B with the same type as the type of the message in question. Therefore, $not\_said(B, K_{AB}^{-1}, (N_A, N_B, B), r_1, 2)$ does not hold and the analysis can not proceed, thus showing that under certain assumptions the protocol is insecure.

This does not mean, of course, that the protocol is insecure in general. There are many ways to avoid the above attack. It is impossible in environments where interleaving protocol runs are not allowed. An attack can easily be avoided by requiring the recipient of the first message to check that the textfield (if it contains the identification of the message's sender) is not equal to the recipient's identification. A different method is to use direction fields for each message, as discussed in [Syv93]. Typing does not help here, though, since all datafields used in the attack are used with respect to their type (a nonce is used as a nonce etc.).

However, there are environments where the attack is possible. The following describes a way a manipulated smartcard reader SCR can authenticate itself to a smartcard SC without performing any encryption, using the above version of the protocol.

$(r_1, 1). SCR \longrightarrow SC : \text{GET\_CHALLENGE}$
$(r_1, 2). SC \longrightarrow SCR : N_{SC}$
$(r_1, 3). SCR \longrightarrow SC : \text{GET\_DATA}$
$(r_1, 4). SC \longrightarrow SCR : ID_{SC}$
$(r_2, 1). SCR \longrightarrow SC : \text{MUTUAL\_AUTHENTICATE}(N_{SC}, ID_{SC})$
$(r_2, 2). SC \longrightarrow SCR : \{K; N_{SC}', N_{SC}, ID_{SC}\}$
$(r_1, 5). SCR \longrightarrow SC : \text{MUTUAL\_AUTHENTICATE}(\{K; N_{SC}', N_{SC}, ID_{SC}\})$
$(r_1, 6). SC \longrightarrow SCR : \text{no error}$

Note that the key K to be used by the smartcard for performing the command MUTUAL_AUTHENTICATE is determined not in its data field but in the parameter P2. (See [7816-4] and [7816-8], respectively, for a detailed description of the smartcard commands). A smartcard is not able to detect the above attack when using the protocol description presented in [9798-2].

7 Conclusions

We have introduced a formalism capable of detecting the possibility of certain reflection and interleaving attacks on authentication protocols using a symmetric algorithm. By applying SG logic we were able to show the weakness of the Neuman-Stubblebine protocol formally. Furthermore, we have shown that due to inaccuracies in the protocol description the three pass authentication protocol 5.2.2 standardized in ISO/IEC 9798-2 is susceptible to an interleaving attack in certain environments. As the same inaccuracies can be found in the description of the two pass authentication protocol 5.1.2 (see [9798-2]), this protocol is equally vulnerable (see [Gue96] for a detailed analysis of this protocol).

The weaknesses of the protocols discussed in this paper are due to the checks that are or are not performed by the principals receiving a message rather than to the messages themselves. The analyses given emphasize the importance of formalizing those checks, which is achieved by using the notion of the type of a message. However, it is desirable to improve the SG logic with respect to determining the set of possibly generated messages. For this a formal language has to be developed by means of which the connection between received and sent messages can be described, thus enabling us to describe precisely how a key to be used is determined, which datafields of the received messages are used in the next message to be sent etc. Based upon this a formal protocol description can be provided by means of which the sets of possibly generated messages can be determined formally.

References

[AbTu91] Abadi, M., Tuttle, M.R.: A Semantics for a Logic of Authentication. Proceedings of the Tenth Annual ACM Symposium on Principles of Distributed Computing, Montreal, Canada, August 1991, pp. 201-216.

[BAN89] Burrows, M., Abadi, M., Needham, R.: A Logic of Authentication. Report 39 Digital Systems Research Center, Palo Alto, California, 1989.

[Ca93] Carlsen, U.: Using Logics to Detect Implementation-Dependent Flaws. Ninth Annual Computer Security Applications Conference, De. 6-10, 1993, Orlando, Florida, pp. 64-73.

[X.509-87] CCITT Recommandation X.509 - ISO/IEC 9594/8 (1987)

[DeSa81] Denning, D.E., Sacco, G.M.: Timestamps in Key Distribution Protocols. CACM Vol. 24, No. 8, August 1982, pp. 533-536.

[ElGa85] ElGamal, T.,: A Public Key Cryptosystem and a Signature Scheme based on Discrete Logarithms; IEEE Transactions on Information Theory, Vol.31, Nr.4, July 1985, S.469-472

[GaSn91] Gaarder, K., Snekkenes, E.: Applying a Formal Analysis Technique to the CCITT X.509 Strong Two-Way Authentication Protocol. J. Cryptology, Vol. 3, No. 2 (1991) 81–98.

[GNY90] Gong, L., Needham, R., Yahalom, R.: Reasoning about Belief in Cryptographic Protocols. Proc. 1990 IEEE Symp. on Security and Privacy (Oakland, California), pp. 234-248.

[Gue96] Güergens, S.: A Formal Analysis Technique for Authentication Protocols. Proceedings of Pragocrypt 1996.

[7816-4] ISO/IEC 7816-4: 1995(E) "Information technology - Identification cards - Integrated circuit(s) cards with contacts - Part 4: Inter-industry commands for interchange"

[7816-8] WD 7816-8: 1995(E) "Information technology - Identification cards - Integrated circuit(s) cards with contacts - Part 8: Security related interindustry commandsx"

[9798-2] ISO/IEC 9798-2: 1994(E) "Information technology - Security techniques - Entity authentication - Part 2: Mechanisms using encipherment algorithms"

[MSN94] Mathuria, A., Safavi-Naini, R, Nickolas, P.: Some Remarks on the Logic of Gong, Needham and Yahalom. Proceedings of the International Computer Symposium 1994, Dec. 12-15, NCTU, Hsinchu, Taiwan, R.O.C, Vol.1, pp. 303-308

[NeSch78] Needham, R. M., Schroeder, M. D.:Authentication Revisited. Operating Systems Review, Vol. 21, No. 12, Decemter 1978, pp. 993-999.

[NeSt93] Neuman, B.C., Stubblebine, S.G.: A note on the use of timestamps as Nonces. Operating Systems Rev. 27 (2), 1993, pp. 10-14.

[RSA78] Rivest, R. L., Shamir, A., Adleman, L. A.: A method for obtaining digital signatures and public-key cryptosystems; Communications of the ACM, Vol.21, Nr.2, 1978, S.120-126.

[Syv93] Syverson, P.: On Key Distribution Protocols for Repeated Authentication; Operating Systems Review, vol. 27, no. 4, October 1993, pp. 24-30.

[vO93] Van Oorschot, P.C.: Extending Cryptographic Logics of Belief to Key Agreement Protocols. 1st ACM Conference on Computer and Communications Security, Nov. 3-5 1993, Fairfax, Virginia, pp. 232-243.

How to Convert any Digital Signature Scheme into a Group Signature Scheme

Holger Petersen[1]

Laboratoire d'informatique
Ecole Normale Supérieure
45, rue d'Ulm, F-75005 Paris
E-Mail: hpetersen@geocities.com

Abstract. Group signatures, introduced by Chaum and van Heijst, allow members of a group to sign messages anonymously on behalf of the group. In this paper we show, how any digital signature scheme can be converted into a group signature scheme. This solves an open problem posed by Chaum and van Heijst. To obtain this result, the encrypted identity of the signer is signed together with the message. The correctness of the encryption is ensured by an *indirect disclosure proof* that proves that the group center has the capability to decrypt the signers identity later.

Furthermore, an efficient extension to the general case of a (t, n) threshold group signature scheme with $2 \leq t < n$ anonymous signers is described. Finally, we discuss some extensions of both schemes, like selective and total conversion into ordinary digital signatures.

1 Introduction

Group signatures allow members of a group to sign messages on behalf of the group, such that

1. the verifier of the signature can check, that the signature is a valid group signature but cannot discover which group member made it,
2. in case of a dispute, either a trusted authority, called *group center*, or the group members together can identify the signer.

Group signature schemes have first been proposed by Chaum and van Heijst [ChHe91]. Their schemes have been improved by Chen and Pedersen [ChPe94]. They describe a general technique to identify the signer by double-signing the message. More recently there were several new proposals promising nice computational properties [KiPW96] which have unfortunately been broken by Michels [Mich96]. This recent publication, however, proposes to add the property of *total conversion* to the group signature. In this context it is also possible to consider the *selective conversion*, which gives the following concepts:

---

[1] This work has been supported by a postdoctoral fellowship of the NATO Science Committee disseminated by the DAAD. His current address is r^3 *security engineering, Zürichstrasse 151, CH-8607 Aathal, E-Mail: petersen@r3.ch*

- *Total convertible group signature:* The release of some secret information of the signer turns *all* of his group signatures into ordinary digital signatures.
- *Selective convertible group signature:* By releasing some secret information, the signer turns *any* of his group signatures into an ordinary digital signature.

The difference between "selective conversion" and "identification by the group center" is the fact, if the signer identifies himself voluntarily or not.

It is possible to generalize the concept of group signatures, such that any coalition of t out of n anonymous group members sign a message together. Then the verifier checks that the signature has been generated by t members of a group without knowing which ones. Their identities are revocable in the case of a dispute by the group center (or the group members). More formally, group signature schemes should satisfy these properties:

- *Authorization:* Only authorized group members can generate group signatures.
- *Unforgeability:* A group signature is not forgeable[2] by any unauthorized (group of) person(s).
- *Untraceability:* It is impossible[2] to find out which member (coalition of members) signed a message.
- *Unlinkability:* It is impossible[2] to find out which messages are signed by the same member (coalition of members).
- *Revocability:* In case of a dispute, the group center or a coalition of group members can identify the signer (coalition of signers).
- *No framing:* No coalition of group members nor the group center can falsely accuse[2] a group member (coalition of members) having signed a message that he (they) didn't sign.
- *Efficiency:* The group center is not involved in signing and verification of signatures. Otherwise he could identify the signer trivially.

The *untraceability* property might be strengthend by the requirement, that it should also be impossible to find out, *how many* of the group members (or more generally which access structure) signed a message. This might be useful, if the verifier knows all possible access structures in a group that are able to sign and is willing to accept any of them. For practical issues it is more likely that he doesn't have this information and wants to be sure about how many persons in the group signed a message together. Therefore, we don't consider this strengthend notation in the following.

Chaum and van Heijst stated the following as an open problem for the design of group signature [ChHe91]:

- "Is it possible to make digital group signatures (with computational anonymity) other than by using [FiSh86] on undeniable signatures ?"

There has been a first attempt to solve this problem in the insecure scheme of [KiPW96] which was broken in [Mich96]. We explain in this paper how to solve

---

[2] usually under some complexity measure, e.g. polynomial time computation

this problem by allowing the use of *any* secure, memoryless digital signature scheme, that follows the description of [GoMR88].

1.1 Related work

The following table compares the properties of some of the proposals in [ChHe91] and [ChPe94] with the new scheme. For a complete overview of the properties of all previous schemes see [KiPW96].

Property	[ChHe91] scheme 1	[ChHe91] scheme 4	[ChPe94] scheme 2	new scheme
Type of signature	any	undeniable	undeniable	*any*
Signature based on assumption	any	DLOG	DLOG	any
Framing based on assumption	any	DLOG	DLOG	DLOG
Framing by group center	possible	possible	possible	*impossible*
Anonymity based on assumption	unconditional	DLOG	DLOG	DH
Identification of the signer by	group center	coalition of members	coalition of members	coalition of members
Effort of identification	linear in $|G|$	linear in $|G|$	linear in $|G|$	*constant*
Group key (GK)	user keys	user keys	user keys	*common*
Length of public GK	linear in $|G|$	linear in $|G|$	linear in $|G|$	*fixed*
Number of computat.	independent	linear in $|G|$	linear in $|G|$	linear in $|G|$
Length of signature	independent	linear in $|G|$	linear in $|G|$	linear in $|G|$
Inclusion of new group members	no	yes	yes	yes
Efficient extension for $2 \leq t \leq n$ signers	no	no	no	*yes*

One of the advantages of our schemes compared to the existing ones is the *independent choice* of the signature scheme and the signer identification mechanism. Therefore *any* signature scheme can be used and the size of the public key is *fixed*. Also the identification of the signer is achieved by a *constant* effort and framing of a honest group member by the group center is *impossible*. Another advantage is the natural, efficient extension of our scheme for the case of $t > 1$ anonymous signers, which isn't known for any published scheme.

1.2 Overview of the results

We present a general description of the new scheme, that uses an *indirect disclosure proof* [3] as important tool. Hereafter we give a concrete realization of this

---

[3] The terminology is chosen similar to [FrTY96], as their *indirect discourse proof* has comparable characteristics. We use a new name for clear distinction.

proof to complete the description. Then we adapt our results to the general case of a (t, n) threshold group signature scheme. Finally, we present the *selective* and *total* conversion of our schemes into ordinary digital signatures and show how to share the encrypted identities among the group members such that for some $k \geq 1$ any coalition of k group members can identify the signer.

2   A generic group signature scheme

We describe the generic scheme, which can be set up using any digital signature scheme. As tools we use a cryptographic hash function h, a probabilistic public key cryptosystem $(\mathcal{E}, \mathcal{D})$ and an indirect disclosure proof (IDP) V which will be specified in section 4. The underlying signature scheme $(\mathcal{G}, \mathcal{S}, \mathcal{V})$ with key generation algorithm $\mathcal{G}$, signature algorithm S and verification algorithm $\mathcal{V}$ is specified with respect to [GoMR88]. In contrast to the schemes in [ChHe91, ChPe94], we propose a scheme where a *common key pair* is used and shared among all members of the group. We assume, that the group G consist of n members $U_1, \ldots, U_n$ with identities $ID_1, \ldots, ID_n$. These identities are an integral part of the description of the group.

1. *Initialisation:* The certification authority generates large primes p, q, with $q|(p-1)$, generators α and β of order q and a suitable collision resistant hash function h. Additionally, it chooses a probabilistic encryption scheme with encryption function $\mathcal{E}$.
2. *Key generation:* The common group keypair (x_G, y_G) is generated, certified and distributed confidentially to all group members. Additionally, the group center Z generates its own keypair (x_Z, y_Z). The public key y_Z is certified and published to all users.
3. *Signature generation:* User $U \in G$ with identity ID_U generates a signature on the message m by computing the ciphertext $c := \mathcal{E}(y_Z, ID_U)$ and signing the message $\bar{m} := h(m, c)$ using the signature algorithm $\mathcal{S}$, that is $\sigma := \mathcal{S}(x_G, \bar{m})$. Furthermore he generates a message-dependent indirect disclosure proof $V(c)$, which proves that c encrypts *exactly* his identity ID_U under the secret key x_Z, without revealing ID_U. The signed message consists of the tuple $(m, c, \sigma, V(c))$.
4. *Signature verification:* The verification of the group signature consists of two parts: First the digital signature σ on the message $h(m, c)$ is checked using $\mathcal{V}$. Second, the indirect disclosure proof $V(c)$ is verified to ensure the correct generation of the ciphertext c.
5. *Identification of the signer:* The group center Z decrypts the ciphertext c using its secret key x_Z. It obtains ID_U and identifies the signer.

Security considerations

We consider the general security aspects. The security of the scheme relies on the security of the signature scheme $(\mathcal{S}, \mathcal{V})$, the probabilistic encryption scheme $(\mathcal{E}, \mathcal{D})$, the hash function h and the indirect disclosure proof $V(c)$.

1. The *security against forgery* of a signature relies on the underlying signature scheme. The group center Z and the other group members can't create a valid signature on behalf of user U, as they can't generate the IDP $V(c)$. As the proof $V(c)$ is message-dependent it is impossible to replay a previous proof $V(\tilde{c})$ and combine it with a fresh message.

2. The ciphertext c has to be encrypted using a probabilistic encryption scheme $(\mathcal{E}, \mathcal{D})$. Otherwise the encryption of identity ID_U might be tested by successively encrypting the identities of all group members and comparing the results with c.

3. The probabilistic encryption scheme guarantees the *unlinkability* of different signed messages of the same signer, as the encryption of the identity ID_U results in different ciphertexts c each time a message m is signed.

4. *Privacy protection* of a signer relies on the cryptographic strength of the encryption function $\mathcal{E}$, as his identity is revealable by decrypting c.

5. *Framing* of a group member depends on the ability of other members to generate a valid IDP $V(c)$. It will be discussed in section 4.

6. The correct *identification* of the signer depends first on the unforgeability of the indirect disclosure proof $V(c)$ which will be discussed in section 4. Second it relies on the collision resistance of the hash function h, that the ciphertext c in a signed message $(h(m, c), \sigma)$ can't be replaced by any other ciphertext c' satisfying $h(m, c) = h(m, c')$.

3 Cryptographic tools

As proof of knowledge of a representation we use a proof of Okamoto [Okam92], as proof of equality of two discrete logarithms, we employ the proof described in [ChPe92].

Definition 1 (Proof$_{REP}$) *A message-dependent proof of knowledge of the representation of y to the base (α, β) consists of a signature on the message $m||\alpha||\beta||y$ with respect to the public key y and the base (α, β). It is denoted as $(r, s_1, s_2) := Proof_{REP}(m, \alpha, \beta, y)$, where $r := H(m||\alpha||\beta||y||\alpha^{s_1} \cdot \beta^{s_2} \cdot y^r)$.*

The prover Alice, who knows a representation $y := \alpha^{x_1} \cdot \beta^{x_2} \pmod{p}$ can construct the proof. She chooses two random numbers $k_1, k_2 \in_R \mathbf{Z}_q$ and computes $r := H(m||\alpha||\beta||y||\alpha^{k_1} \cdot \beta^{k_2})$. Then she calculates $s_1 := k_1 - r \cdot x_1 \pmod{q}$ and $s_2 := k_2 - r \cdot x_2 \pmod{q}$. It is assumed that this proof doesn't leak any information about x_1 and x_2.

Definition 2 (Proof$_{LOGEQ}$) *A message-dependent proof of equality of the discrete logarithm of y to the base α and z to the base β is a tuple $(r, s) := Proof_{LOGEQ}(m, \alpha, y, \beta, z)$, where $r := H(m||\alpha||y||\beta||z||\alpha^s \cdot y^r||\beta^s \cdot z^r)$.*

This proof can be obtained, if and only if the prover knows the discrete logarithms $\log_\alpha(y)$ and $\log_\beta(z)$ and they are both equal. To construct the proof, the prover chooses $k \in_R \mathbf{Z}_q$ and calculates $r := H(m||\alpha||y||\beta||z||\alpha^k||\beta^k)$ and $s := k - r \cdot x \pmod{q}$.

Third, we need a proof of knowledge of *one* out of n discrete logarithms which is equal to another logarithm. An efficient proof of <u>or</u> has been proposed by [Scho93] and described in [ChPe94], which is now combined with Proof_{LOGEQ}.

Definition 3 ($\text{Proof}_{\exists LOGEQ}$) *A message-dependent proof of knowledge that the logarithm of one of the y_i, $i \in [1 : n]$, to the base α is equal the discrete logarithm of z to the base β, without revealing for which y_i this relation holds, consists of a signature on the message $m||y_1|| \cdots ||y_n||\alpha||z||\beta$ with respect to the public values y_i, z and the bases α, β. It is denoted as $V := \text{Proof}_{\exists LOGEQ}(m, y_1, \ldots, y_n, \alpha, z, \beta)$, where $V := (c, c_1, \ldots, c_n, s_1, \ldots, s_n)$ and*

$$c := H\left(m||y_1||...||y_n||\alpha||z||\beta||\alpha^{s_1}y_1^{-c_1}||...||\alpha^{s_n}y_n^{-c_n}||\beta^{s_1}z^{-c_1}||...||\beta^{s_n}z^{-c_n}\right) = \sum_{i=1}^{n} c_i.$$

To simplify notations, we *assume in the following* that the relation holds for $i = 1$, e.g. $x_1 = \log_\alpha(y_1) \equiv \log_\beta(z) \pmod{q}$. The interactive proof is described in the following table.

1. Alice chooses random $k_i \in_R \mathbf{Z}_q^*, i \in [1 : n], c_j \in_R \mathbf{Z}_q^*, j \in [2 : n]$ and computes
  $-$ $r_1 := \alpha^{k_1} \pmod{p}$, $r_i := \alpha^{k_i} \cdot y_i^{-c_i} \pmod{p}$ for $i \in [2 : n]$,
  $-$ $t_1 := \beta^{k_1} \pmod{p}$, $t_i := \beta^{k_i} \cdot z^{-c_i} \pmod{p}$ for $i \in [2 : n]$.
  She sends all values $(r_1, \ldots, r_n, t_1, \ldots, t_n)$ to the verifier Bob.
 2. Bob chooses a random challenge $c \in_R \mathbf{Z}_q^*$ and sends it to Alice.
 3. Alice first calculates $c_1 := c - \sum_{i=2}^{n} c_i \pmod{q}$, then $s_1 := x_1 \cdot c_1 + k_1 \pmod{q}$
  and $s_i := k_i$ for $i \in [2 : n]$ and sends the tuple $(c_1, \ldots, c_n, s_1, \ldots, s_n)$ to Bob.
 4. Bob checks that $c = \sum_{i=1}^{n} c_i$ and that for all $i \in [1 : n]$:
 $$\alpha^{s_i} \equiv y_i^{c_i} \cdot r_i \pmod{p} \text{ and } \beta^{s_i} \equiv z^{c_i} \cdot t_i \pmod{p}.$$

Table 1. Honest verifier Zero-Knowledge-Proof $\text{Proof}_{\exists LOGEQ}$

Theorem 1 (following [Scho93]) *The protocol in table 1 is a* witness indistinguishable *proof of knowledge of x_i satisfying*

$$y_i \equiv \alpha^{x_i} \pmod{p} \text{ and } z \equiv \beta^{x_i} \pmod{p} \text{ for some } i \in [1 : n].$$

Proof 1:

1. *Completeness:* The following equivalences hold by construction of the proof:

$$i = 1 : \alpha^{s_1} \equiv \alpha^{x_1 c_1 + k_1} \equiv y_1^{c_1} r_1 \pmod{p}, \quad \beta^{s_1} \equiv \beta^{x_1 c_1 + k_1} \equiv z^{c_1} t_1 \pmod{p},$$
$$i \geq 2 : \alpha^{s_i} \equiv \alpha^{k_i} \equiv y_i^{c_i} \alpha^{k_i} y_i^{-c_i} \equiv y_i^{c_i} \cdot r_i, \quad \beta^{s_i} \equiv z^{c_i} \beta^{k_i} z^{-c_i} \equiv z^{c_i} t_i \pmod{p}.$$

2. *Soundness:* Suppose, that Alice is able to answer two different challenges c_1 and c_2 without knowing any of the secret keys $x_1, \ldots, x_n$ after she has transmitted the values $(r_1, \ldots, r_n, t_1, \ldots, t_n)$, such that for all $i \in [1:n], j \in [1:2]$:

$$\alpha^{s_{i,j}} \equiv y_i^{c_{i,j}} \cdot r_i \pmod{p}, \quad \beta^{s_{i,j}} \equiv z^{c_{i,j}} \cdot t_i \pmod{p} \quad \text{and} \quad c_j = \sum_{i=1}^{n} c_{i,j}.$$

Then there exists at least one pair of tuples $(s_{i,1}, c_{i,1})$ and $(s_{i,2}, c_{i,2})$ such that $c_{i,1} \neq c_{i,2}$ and $\alpha^{s_{i,1}} \equiv y_i^{c_{i,1}} \cdot r_i \pmod{p}$, $\alpha^{s_{i,2}} \equiv y_i^{c_{i,2}} \cdot r_i \pmod{p}$. Thus $\alpha^{s_{i,1} - s_{i,2}} \equiv y_i^{c_{i,1} - c_{i,2}} \pmod{p}$ for some i, which implies

$$y_i \equiv \alpha^{\frac{s_{i,1} - s_{i,2}}{c_{i,1} - c_{i,2}}} \pmod{p} \quad \text{or} \quad x_i \equiv \frac{s_{i,1} - s_{i,2}}{c_{i,1} - c_{i,2}} \pmod{q}.$$

Thus Alice knows x_i which contradicts the assumption above.

3. *Witness indistinguishable:* The verifier cannot determine which of the x_i's the prover knows. The transcripts (r_i, c_i, s_i) and (t_i, c_i, s_i) have all the same distribution as in a single Schnorr signature, except that they satisfy the additional property $\sum_{i=1}^{n} c_i = c$ which doesn't help the verifier. Thus the protocol reveals no information about which of the n witnesses Alice knows. Even a distinguisher, who knows the possible witnesses, cannot tell witch witness she knows. Thus the protocol is witness indistinguishable in the sense of [FeSh90]. □

The interactive proof in table 1 is transformable into a non-interactive proof using standard techniques [FiSh86]. This results in

$$Proof_{\exists LOGEQ}(y_1, \ldots, y_n, \alpha, z, \beta) := (m, c, c_1, \ldots, c_n, s_1, \ldots, s_n), \quad \text{where}$$

$$c := H\left(m||y_1||\ldots||y_n||\alpha||z||\beta||\alpha^{s_1} y_1^{-c_1}||\ldots||\alpha^{s_n} y_n^{-c_n}||\beta^{s_1} z^{-c_1}||\ldots||\beta^{s_n} z^{-c_n}\right) = \sum_{i=1}^{n} c_i.$$

As an instance for $n = 1$ we obtain the proof of equality of two discrete logarithms [ChPe92]. Using these proofs, we are now ready to describe the indirect disclosure proof $V(c)$.

4 Indirect Disclosure Proof

We develop an indirect disclosure proof which demonstrates that the ciphertext $c := \mathcal{E}(y_Z, ID_i)$ encrypts exactly the signers identity ID_i under the group centers public key y_Z. It consists of two parts:

1. c encrypts *at most* the identity of the signer.
2. c encrypts *at least* one of the identities ID_i, $i \in [1:n]$.

If both parts are proven, c has to encrypt exactly the identity of the signer. The realization depends on the choice of the probabilistic encryption scheme $(\mathcal{E}, \mathcal{D})$ and thus the structure of the ciphertext c. In the following, we focus on the ElGamal encryption scheme as an example [ElGa85], but in principal it can be developed using any other probabilistic cryptosystem like e.g. the randomized RSA scheme. To avoid, that the signer cheats by encrypting the identity of another group member, each ID_i is generated from a secret pre-image u_i using a one-way function f, that is $ID_i := f(u_i)$. The proof of knowledge of this pre-image (without revealing it) is the first part of the proof. The pre-image can be seen as the user's *secret key* and his identity ID_i as corresponding *public key*.

To show, that at most one identity is encrypted in c, we use Proof$_{REP}$ – to prove that at least one identity is embedded, we use Proof$_{\exists LOGEQ}$. Thus, the indirect disclosure proof is a concatenation of these proofs. To simplify the notations, we assume, that *signer Alice* generates the group signature. The initialization and key generation are as in chapter 2.

1. *Initialization:* Choice of primes p, q, generators α, β and hash function h.
2. *Key generation:* Besides the generation of the group keypair (x_G, y_G) and the center's pair (x_Z, y_Z), the signer Alice chooses a secret value $u_A \in_R \mathbf{Z}_q$ and calculates $ID_A := \beta^{u_A} \pmod{p}$, which is published as her identity.
3. *Encryption of the identity:* Alice chooses a random $k_c \in_R \mathbf{Z}_q$, computes $c_1 := \alpha^{k_c} \pmod{p}$ and $c_2 := y_Z^{k_c} \cdot ID_A \pmod{p}$. The tuple $c := (c_1, c_2)$ is the ciphertext of ID_A which is hashed together with m before signing.
4. *Indirect disclosure proof:* Alice generates $U(c_2) :=$Proof$_{REP}(m, y_Z, \beta, c_2)$, to assure that she knows a representation of c_2 to the base (y_Z, β) and $W(c_1, c_2) :=$Proof$_{\exists LOGEQ}(m, c_2/ID_1, \ldots, c_2/ID_n, y_Z, c_1, \alpha)$, to assure, that one of the logarithms of $d_i := c_2/ID_i \pmod{p}$ to the base y_Z is equal to the discrete logarithm $\log_\alpha(c_1) \equiv k_c \pmod{q}$. The indirect disclosure proof $V(c_1, c_2)$ is the concatenation of $U(c_2)$ and $W(c_1, c_2)$. It is distributed together with the signed message (m, c, σ).
5. *Identification of the signer:* As $c_1 \equiv \alpha^{k_c} \pmod{p}$, $c_2 \equiv y_Z^{k_c} \cdot ID_A \pmod{p}$ and $y_Z \equiv \alpha^{x_Z} \pmod{p}$ by definition, the group center obtains Alice's identity by $c_1^{-x_Z} \cdot c_2 \equiv \alpha^{-x_Z k_c} \cdot y_Z^{k_c} \cdot ID_A \equiv ID_A \pmod{p}$. To prove that ID_A is the correct identity, it also generates Proof$_{LOGEQ}(m, \alpha, c_1, y_Z, c_2/ID_A)$.

4.1 Security

Theorem 2 *Assume, that the representation problem is hard and that Alice doesn't know $\log_{y_Z}(ID_i)$ for any user's ID_i. Then Proof$_{REP}(m, y_Z, \beta, c_2)$ and Proof$_{\exists LOGEQ}(m, c_2/ID_1, \ldots, c_2/ID_n, y_Z, c_1, \alpha)$ assure, that (c_1, c_2) encrypts an identity ID for which Alice knows the pre-image u, i.e. $ID := \beta^u \pmod{p}$.*

Proof 2: Proof$_{REP}(m, y_Z, \beta, c_2)$ assures that

$$c_2 := y_Z^{v_1} \cdot \beta^{v_2} \pmod{p} \text{ for any } v_1, v_2 \in_R \mathbf{Z}_q. \tag{1}$$

$Proof_{3LOGEQ}(m, c_2/ID_1, \ldots, c_2/ID_n, y_Z, c_1, \alpha)$ assures, that one of the logarithms of c_2/ID_i (mod p) to the base y_Z is equal to the discrete logarithm $\log_\alpha(c_1) \equiv k_c$ (mod q). Thus

$$c_2 \equiv y_Z^{k_c} \cdot ID_i \pmod{p} \tag{2}$$

Therefore it follows: $v_1 \equiv k_c$ and $\beta^{v_2} \equiv ID_i$ (mod p). Thus v_2 must be equal to u_A and $ID_i = ID_A$. Otherwise, Alice would be aware of the discrete logarithm of another user's identity ID_i to the base β (or got knowledge of it by the user U_i himself). $\square$

Theorem 3 Privacy protection *of Alice against any other person than the group center is based on the Diffie-Hellman problem.*

Proof 3: Alice's identity is ElGamal encrypted which is equivalent to the Diffie-Hellman problem, if the secret values k_c and x_Z are unknown [ElGa85]. $\square$

Theorem 4 Framing *of Alice by other group members is based on the discrete logarithm problem.*

Proof 4: To frame Alice, one has to encrypt her identity in (c_1, c_2) and generate a valid proof $V(c_1, c_2)$. This can only be done, if the pre-image u_A of ID_A is known, as shown in theorem 2, which is equivalent to compute the discrete logarithm $\log_\beta(ID_A)$. $\square$

It has been proven, why the identity of the signer is revealed if the two proofs are given correctly. Now we show the opposite.

Theorem 5 *The identification fails if at least one of the two proofs fails.*

Proof 5: We show this for both proofs separately:

- If only $Proof_{3LOGEQ}$ is given, Alice can cheat by generating $c_2 := y_Z^k \cdot ID_i$ (mod p) with $ID_i \neq ID_A$, as she is not obliged to show that she knows the preimage of ID_i to base β. Thus the group center will identify another group member.
- If only $Proof_{REP}$ is given, Alice can cheat by choosing $c_2 := y_Z^{v_1} \cdot \beta^{v_2}$ (mod p) with random $v_1, v_2 \in_R \mathbf{Z}_q$. As with overwhelming probability $\beta^{v_2} \neq ID_i$ for any i and $v_1 \neq k$, no group member can be identified by the group center. $\square$

Remark 1 *The assumption that Alice doesn't know the discrete logarithm of any ID_i to the base y_Z can be dropped, if we substitute $Proof_{REP}$ by $(r, s_1, s_2) := Proof_{REP+EQ_1}(m, y_Z, \beta, c_2, \alpha, c_1)$. It demonstrates, that the representation of c_2 to base (y_Z, β) is known and that the exponent of y_Z in the representation is equal to the discrete logarithm of c_1 to base α. Thus it shows, that c_2 must have the structure $c_2 \equiv y_Z^{k_c} \cdot \beta^{v_2}$ (mod p).*

The proof is obtained by calculating $r := H(m||y_Z||\beta||c_2||\alpha||c_1||y_Z^{k_1} \cdot \beta^{k_2}||\alpha^{k_1})$ with $k_1, k_2 \in_R \mathbf{Z}_q$, $s_1 := k_1 - r \cdot k_c$ (mod q) and $s_2 := k_2 - r \cdot u_A$ (mod q). It is checked by verifying $r := H(m||y_Z||\beta||c_2||\alpha||c_1||y_Z^{s_1}\beta^{s_2}c_2^r||\alpha^{s_1}c_1^r)$.

4.2 Efficiency

The indirect disclosure proof V needs the transmission of $2n+1$ values of length $|q|$ bit. Additionally, σ and the ciphertext (c_1, c_2) of length $|p| + |q|$ bit have to be transmitted. If σ is of length $|p| + |q|$ bit, the *signed message* is of total length $(2n+3) \cdot |q| + 2 \cdot |p|$ bit. This is about half the size of the schemes in [ChPe94], which need $2 \cdot (2n+1) \cdot |q| + |p|$ due to double signing. Another advantage is, that the identification of the signer is done by *one* single exponentiation in our scheme, which needed at least $n := |G|$ exponentiations in all former schemes with computational anonymity.

5 General threshold group signatures

In general, (t, n) threshold group signatures allow a subset of t members of a group to sign messages on behalf of the group such that

1. the verifier of a signature can check that it is a valid group signature generated by t members. He is unable to discover which group members signed the message.
2. In the case of a dispute, either the group members together or a trusted authority (the group center) can identify the t signers.

The scheme is an extension of the basic group signature scheme for a single signer $(t = 1)$. As before any threshold signature scheme can be chosen as underlying signature scheme. For example, the threshold signature scheme of Park and Kurosawa [PaKu96] or the scheme presented by Gennaro et al. [GJKR96] might be used. The latter offers the advantage that it tolerates up to $n/4$ malicious faulty signers that generate incorrect partial signatures. More general, it is also possible to use a signature scheme that allows an arbitrary access structure as proposed in [CrDS94].

The secret group key x_G is *shared* among the n members, such that any t of them can sign a message but less than t can't generate a valid signature corresponding to the public group key y_G. To simplify the notations, we assume that the t members $U_1, \ldots, U_t$ of G sign a message m together.[4] To allow the *identification* of the t signers U_i, each of them encrypts his identity ID_i in the same way as in the basic scheme. The protocol consists of the following steps:

1. The t signers $U_1, \ldots, U_t$ agree upon a message m to be signed.
2. Each signer $U_i, i \in [1 : t]$ computes the ciphertext $c_i := (c_{i,1}, c_{i,2}) = \mathcal{E}(y_Z, ID_i)$ and the proof $V(c_i)$. These values are broadcasted to all signers. Additionally he computes the message-dependent proof $W_i(c_{i,1}, c_{i,2}) := \text{Proof}_{LOGEQ}(m, y_Z, c_{i,2}/ID_i, \alpha, c_{i,1})$ which is send *confidentially* to the participating signers. The proof demonstrates that $c_{i,2}$ encrypts the identity ID_i of signer U_i. This allows each signer to check, that the others encrypt their proper identity in c_i.

---

[4] in general, we have to apply a permutation to the user indices $1, \ldots t$.

3. Each signer U_i checks the proofs W_j, $j \neq i$, and calculates the message $\bar{m} := h(m, c_1, \ldots, c_t)$.
4. The signers calculate the (t, n) threshold signature on the message $\bar{m}$. The signed message consists of the tuple $(m, c_1, \ldots, c_t, \sigma, V(c_1), \ldots, V(c_t))$.

For the security considerations, we make several reasonable assumptions, which are very likely to be satisfied in almost any application of group signatures.

Assumption 1 *Each signer agrees on signing the message m before generating $c_i := (c_{1,i}, c_{2,i})$ and the message-dependent proof $V(c_i)$.*

If this assumption holds, it is not possible that any signer who doesn't want to sign the message m is forced to sign a message against his own will.

Assumption 2 *Each signer U_i checks all proofs W_j, $j \neq i$, before signing $\bar{m}$.*

This assumption is even demanded in step 3 of the protocol. If it holds (e.g. no signer simplifies his protocol by omitting step 3), then all t signers know each other and also that the others have embedded their correct identities into the signed message $\bar{m}$.

Assumption 3 *At least one of the t signers is honest.*

This assumption ensures that the t ciphertexts c_i correspond to *different* identities and that for example no identity is encrypted several times (to let the verifier believe, that more group members signed than actually did). It is not possible for a coalition of less than t signers (especially not for a single dishonest signer) to generate t encrypted identities and sign the corresponding message thereafter, as they are unable to generate a correct threshold signature from their shares of the secret group key x_G.

6 Convertible group signatures

An extension of the above schemes is the integration of a conversion mechanism, which allows the signer(s) to convert a group signature into an ordinary signature. To obtain a *selective* or *total convertible* signature scheme from one of the above group signature schemes, one must remark that the common group key x_G is known to all group members (or can be used by any t of them in the second scheme). Thus a digital signature generated under this key can't be related to any individual signer. Therefore the authenticity of the converted signature has to rely on the indirect disclosure proof $V(c_1, c_2)$ and some additional published information. The idea is to release a proof that permits to check whose identity is encrypted in (c_1, c_2). [5]

---

[5] In the case of general (t, n) threshold group signatures, this results in a new type of *partially convertible* group signatures, if only some of the signers reveal their identities and others do not.

6.1 Selective convertible scheme

To selectively convert a group signature $(m, c, \sigma, V(c))$ signer Alice releases her identity ID_A together with a proof that the ciphertext $c := (c_1, c_2)$ encrypts this. This is done by $U(c_1, c_2) := \text{Proof}_{LOGEQ}(m, y_Z, c_2/ID_A, \alpha, c_1)$ which proves that $\log_{y_Z}(c_2/ID_A) = \log_\alpha(c_1) \pmod{q}$. To obtain this proof, one needs the knowledge of the secret value k_c which is only known by Alice. The security analysis is skipped due to space limitations.

6.2 Total convertible scheme

To obtain a total convertible group signature from the basic group signature scheme (as well as from the general threshold scheme) is not straightforward, as the user's identity is randomly encrypted with each signature. Thus not all signatures can be converted by releasing a single piece of information. To overcome this problem we present a *general construction* to obtain a total convertible scheme from a selective convertible one.

The signer Alice chooses a symmetric encryption scheme (E, D) and a symmetric secret key K_A. For each group signature $(m, \sigma, c, V(c_1, c_2))$ she computes the proof $U(c_1, c_2)$ as in the selective convertible scheme and encrypts it together with her identity as $c_3 := E_{K_A}(U(c_1, c_2), ID_A)$. The ciphertext c_3 is published as part of the group signature.

In the case of a *total conversion* of all group signatures, Alice publishes ID_A and the symmetric key K_A, such that any verifier can decrypt all values c_3, check that ID_A is properly encrypted inside and verify $U(c_1, c_2)$ as in the selective convertible scheme to ensure the authenticity of the signature.

7 Shared identification of signers

A further extension of the schemes is the shared identification of the signer(s) by any subset of k group members. In the above schemes, the identity of the signer is encrypted using the public key y_Z of the group center. If the members of the group verifiable share a revocation keypair (x_R, y_R), as described in the (k, n) threshold cryptosystem of [Pede91], it is possible, that the signer(s) encrypts his identity using y_R. This allows any k members of the group to reveal his identity. In the computation of the indirect disclosure proof $V(c)$, the public key y_Z has to be substituted by y_R accordingly. $V(c)$ guarantees now, that c encrypts the identity of a group member and that any k members of the group that can compute y_R are able to decrypt it.

8 Conclusion

We demonstrated the design of a group signature scheme based on any conventional digital signature scheme. We were also able to extend this approach to the general case $t > 1$ in an efficient manner. A similar result was achieved very recently and independently by Camenisch [Came97].

In principle, it is possible to use any digital signature concept as underlying signature scheme, as for example designated confirmer signatures, directed signatures or blind signatures, which leads to new group signature concepts with different properties.

Acknowledgements

Thanks to Markus Michels and Berry Schoenmakers for sending their manuscripts [Mich96, Scho93] and Jan Camenisch for early dissemination of his paper [Came97]. We also like to thank Markus Stadler and some participants of the workshop for helpful remarks.

References

[Bran93] S.Brands, "'Untraceable Off-Line Cash in Wallets with Observers"', Lecture Notes in Computer Science 773, Advances in Cryptology: Proc. Crypto '93, Springer, (1994), pp. 302 – 318.

[Came97] J.Camenisch, "'Efficient and Generalized Group signatures"', to appear in Lecture Notes in Computer Science , Advances in Cryptology: Proc. Eurocrypt'97, Springer, (1997), 12 pages.

[ChHe91] D.Chaum, E.van Heijst, "Group signatures", Lecture Notes in Computer Science 547, Advances in Cryptology: Proc. Eurocrypt '91, Springer, (1992), pp. 257–265.

[ChPe92] D.Chaum, T.Pedersen, "Wallet databases with observers", Lecture Notes in Computer Science 740, Advances in Cryptology: Proc. Crypto'92, (1993), pp. 89 – 105.

[ChPe94] L.Chen, T.Pedersen, "New Group Signature Schemes", Lecture Notes in Computer Science 950, Advances in Cryptology: Proc. Eurocrypt'94, Springer, (1995), pp. 163 – 173.

[ChPe95] L.Chen, T.Pedersen, "On the Efficiency of Group Signatures Providing Information-Theoretic Anonymity", Lecture Notes in Computer Science 921, Advances in Cryptology: Proc. Eurocrypt'95, Springer, (1995), pp. 39–49.

[CrDS94] R.Cramer, I.Damgard, B.Schoenmakers, "Proofs of Partial Knowledge and Simplifies Designs of Witness Hiding Protocols", Lecture Notes in Computer Science 839, Advances in Cryptology: Proc. Crypto'94, Springer, (1994), pp. 174 – 187.

[ElGa85] T.ElGamal, "A public key cryptosystem and a signature scheme based on discrete logarithms", IEEE Transactions on Information Theory, Vol. IT-30, No. 4, July, (1985), pp. 469 – 472.

[FeSh90] U.Feige, A.Shamir, "'Witness indistinguishable and witness hiding protocols"', Proc. of the 22nd Annual ACM Symposium on the Theory of Computing, ACM Press, (1990), pp. 416–426.

[FiSh86] A.Fiat, A.Shamir, "How to prove yourself: Practical solutions to identification and signature problems", Advances in Cryptology: Proc. Crypto'86, Lecture Notes in Computer Science 263, Springer, (1987), pp. 186 – 194.

[FrTY96] Y.Frankel, Y.Tsiounis, M.Yung, "' "Indirect Discourse Proofs": Achieving Efficient Fair Off-Line E-Cash"', Lecture Notes in Computer Science 1163, Advances in Cryptology: Proc. Asiacrypt'96, Springer, (1996), pp. 286- 300.

[GJKR96] R.Gennaro, S.Jarecki, H.Krawczyk, T.Rabin, "Robust Threshold DSS Signatures", Advances in Cryptology: Proc. Eurocrypt'96, Lecture Notes in Computer Science 1070, Springer, (1996), pp. 354–371.

[GoMR88] S.Goldwasser, S.Micali, R.Rivest, "A secure digital signature scheme", SIAM Journal on Computing, Vol. 17, 2, (1988), pp. 281 – 308.

[KiPW96] S.J.Kim, S.J.Park, D.H.Won, "Convertible Group Signatures", Lecture Notes in Computer Science 1163, Advances in Cryptology: Proc. Asiacrypt'96, (1996), pp. 311 – 321.

[Mich96] M.Michels, "Comments on some group signature schemes", Technical Report TR-96-03, University of Technology Chemnitz-Zwickau, November, (1996), 4 pages, available at *http://www.tu-chemnitz.de/~hpe/techrep.html.*

[Okam92] T.Okamoto, "Provable secure and practical identification schemes and corresponding signature schemes", Lecture Notes in Computer Science 740, Advances in Cryptology: Proc. Crypto '92, (1993), pp. 31 – 53.

[PaKu96] C.Park, K.Kurosawa, "New ElGamal Type Threshold Digital Signature Scheme", IEICE Trans. Fundamentals. Vol. E79-A, No. 1, January, (1996), pp. 86–93.

[Pede91] T.Pedersen, "'A threshold Cryptosystem without a Trusted Party"', Lecture Notes in Computer Science 547, Advances in Cryptology: Proc. Eurocrypt'91, Springer, (1992), pp. 522 – 526.

[Scho93] B.Schoenmakers, "Efficient proofs of OR", manuscript, (1993), 3 pages.

Threshold Key-Recovery Systems for RSA

Tatsuaki Okamoto

NTT Laboratories
Nippon Telegraph and Telephone Corporation
1-1 Hikarinooka, Yokosuka-shi, Kanagawa-ken, 239 Japan
Email: okamoto@sucaba.isl.ntt.co.jp
Tel: +81-468-59-2511
Fax: +81-468-59-3858

Abstract

Although threshold key-recovery systems for the discrete log based cryptosystems such as the ElGamal scheme have been proposed by Feldman and Pedersen [6, 11, 12], no (practical) threshold key-recovery system for the factoring based cryptosystems such as the RSA scheme has been proposed. [1]

This paper proposes the first (practical) threshold key-recovery systems for the factoring based cryptosystems including the RSA and Rabin schemes. Almost all of the proposed systems are unconditionally secure, since the systems utilize unconditionally secure bit-commitment protocols and unconditionally secure VSS.

1 Introduction

In a *key recovery system*, a customer generates a secret key, split it into many shares, and distributes each share to each trustee. If a legal authority or a customer him/herself requires the key, he/she requests some or all of trustees to recovers the key from their shares.

Here, we call a key recovery system a *threshold key recovery system* if a subset of shareholders with more than or equal to a threshold can recover the key, and any subset with less than the threshold cannot recover the key. This threshold property is very important since a few trustees (computing facilities) might be down in a certain period, faulty, or attacked by an enemy.

In this paper, we focus on a threshold key recovery system for a public-key cryptosystem. The most important problems in realizing such a key recovery system is how to prevent a customer from providing an incorrect/garbage share to a trustee, given the corresponding public key. This problem is very serious for the surveillance by an authority, and also relevant even for the service to a customer in case of loosing a key, since a customer and trustees should clarify the responsibility when they fail in recovering the customer's key.

The key tool for the threshold key recovery is the secret sharing [1, 13], especially the verifiable secret sharing (VSS) [4] in order to solve the above-mentioned problem.

Feldman and Pedersen [6, 11, 12] have proposed very efficient and non-interactive VSS protocols based on the discrete logarithm problem. Therefore, their solutions can be directly applied to the threshold key recovery system for "discrete logarithm" based public-key cryptosystems such as the ElGamal scheme.

---

[1] Micali's fair cryptosystems [8] do not have the threshold property. That is, all share holders must participate to recover a key in his scheme, and the key cannot be recovered even without the help of one shareholder.

Although the RSA scheme is the most popular public-key cryptosystem, no threshold key recovery system for the RSA scheme (and for any other factoring based cryptosystems) has been proposed. Micali proposed a key recovery system based on the RSA scheme (fair cryptosystems) [8], but his schemes do not have the threshold property.

This paper proposes the first threshold key-recovery systems for the factoring based cryptosystems including the RSA and Rabin schemes.

The proposed systems utilize unconditionally secure bit-commitment protocols [5, 10] and unconditionally secure VSS [12]. Therefore, our systems are unconditionally secure in the sense that, the customer releases each trustee no additional information than the public key and share. On the other hand, the correctness of shares (no cheating of customer) depends on the computational intractability, difficulty of the discrete logarithm.

Our systems require the interaction between a customer and trustees and are less efficient than Pedersen's threshold key-recovery systems for the discrete logarithm based cryptosystems, which requires no interaction. However, our systems are still very practical and we can construct non-interactive variants with almost the same efficiency as the original schemes.

This paper is organized as follows: Section 2 introduces some conventional techniques, bit-commitments and VSS. In sections 3 through 5, we propose three threshold key recovery schemes for RSA or Rabin. Section 6 evaluates the proposed schemes in the light of efficiency.

2 Conventions

2.1 Bit Commitments

Unconditionally secure bit commitment schemes play very important roles in many cryptographic protocols [11, 12, 5, 10]. Here, we introduce such a bit-commitment scheme based on the discrete logarithm problem.

Assume that T sets up the commitment scheme and U commits to a number. Finally U proves to T that a value is correctly generated without revealing committed information, by using some protocols to be described later.

To set up the commitment scheme, T generates prime p, where $p - 1$ is divisible by q. T also generates g and h whose orders in the multiplicative group Z_p^* are q.

U commits to an integer $s \in Z_q$ by choosing $r \in_U Z_q$ and computing the commitment

$$BC_g(s, r) = g^s h^r \bmod p,$$

where $r \in_U Z_q$ means that r is selected randomly and uniformly from Z_q. This is called a base-g commitment. A commitment is opened by revealing s and r.

2.2 Pedersen's VSS

Here, we also use the bit-commitment, $BC_g(s, r)$, introduced in the previous subsection. So, we follow the same notations for the bit-commitment. In the following (k, n) VSS protocol, dealer U distributes s to n shareholders, $T_1, \ldots, T_n$, and any k shareholders can recover s.

1. U publishes a commitment to s: $BC_g(s, r)$ for a random integer $r \in_U Z_q$.

2. U uniformly chooses $F_j \in_U Z_q$ for $j = 1, \ldots, k-1$, and let $F(x) = s + F_1 x + \cdots + F_{k-1} x^{k-1}$. U computes $s_i = F(i) \bmod q$ for $i = 1, \ldots, n$. U also uniformly chooses $G_j \in_U Z_q$ for $j = 1, \ldots, k-1$, computes $E_j = BC_g(F_j, G_j)$ for $j = 1, \ldots, k-1$, and broadcasts E_j.

3. Let $G(x) = r + G_1 x + \cdots + G_{k-1} x^{k-1}$, and let $t_i = G(i) \bmod q$ for $i = 1, \ldots, n$. Then U sends (s_i, t_i) secretly to T_i for $i = 1, \ldots, n$.

4. When T_i receives his share (s_i, t_i), he verifies that

$$BC_g(s_i, t_i) = BC_g(s, r)(\prod_{j=1}^{k-1} E_j^{i^j}) \bmod p.$$

5. Any k shareholders, $T_{1_1}, \ldots, T_{i_k}$, can recover s from $s_{1_1}, \ldots, s_{i_k}$ by using the standard interpolation formula [13].

3 Scheme 1 for RSA and Rabin

Before describing Scheme 1 protocol, we will introduce a subprotocol regarding the bit-commitment.

3.1 Bit-Commitment Protocol

This protocol is a variant of [9]. Here, the notations for $(BC_g(s, r), p, q, g, h)$ follow those in Section 2.

Protocol: COMPARE COMMITMENTS with bases of order q between y and y'

Common input: y, y' and (p, q, g, a, h). Here, $\mathrm{Ord}(a) = q$.

What to prove: U knows $(s, r, r') \in Z_q \times Z_q \times Z_q$ such that $y = BC_g(s, r)$, $y' = BC_a(s, r')$.

1. Prover U uniformly chooses $v \in_U Z_q$, $w \in_U Z_q$, and $w' \in_U Z_q$, and calculates

$$u = BC_g(v, w), \quad u' = BC_a(v, w').$$

U sends (u, u') to verifier T.

2. T uniformly selects $\beta \in_U Z_q$ and sends it to U.

3. U calculates

$$z = v + s\beta \bmod q, \quad t = w + r\beta \bmod q, \quad t' = w' + r'\beta \bmod q,$$

and send (z, t, t') to T.

4. T checks whether the following equations hold.

$$BC_g(z, t) = uy^{\beta} \bmod p, \quad BC_a(z, t') = u'y'^{\beta} \bmod p.$$

Note: Non-interactive versions of COMPARE COMMITMENTS protocols employ the standard conversion technique using hash functions [7].

3.2 Scheme 1 protocol

Now we present scheme 1, in which the RSA and Rabin schemes are supposed as public-key cryptosystems. User U generates a pair of secret and public keys of the RSA or Rabin scheme, (P, Q) and $N = PQ$. We assume that $(-1/N) = 1$.

1. U generates two primes P and Q, and calculates $N = PQ$ such that $(-1/N) = 1$.
   U sends N to trustees, $T_i, \ldots, T_n$.
2. U also gives the evidence to the trustees that N consists of two primes. Such a non-interactive proof (evidence) for the purpose is described in [8], and such an interactive proof is described in [2].
3. The trustees generate (p, g, h) such that p is a prime and $N|p - 1$, and $\mathrm{Ord}(g) = \mathrm{Ord}(h) = N$. They send (p, g, h) to U. [2]
   Here, $BC_g(x, r) = g^x h^r \bmod p$.
4. U generates random numbers $x \in_U Z_N$, $r_1 \in_U Z_N$, $r_2 \in_U Z_N$, $r_3 \in_U Z_N$ and calculates a, b, c as follows:

$$a = BC_g(x, r_1),$$

$$b = BC_a(x, r_2) = BC_g(x^2 \bmod N, r_1 x + r_2 \bmod N),$$

$$c = BC_b(x^2 \bmod N, r_3) = BC_g(x^4 \bmod N, r_2 x^2 + r_3 \bmod N).$$

   U sends (a, b, c) and $(x^4 \bmod N, r_2 x^2 + r_3 \bmod N)$ to trustees T.
   U also calculates and sends y to T such that $y^2 \equiv x^4 \pmod N$ and $(y/N) = -1$.
   T checks whether $c = BC_g(x^4 \bmod N, r_2 x^2 + r_3 \bmod N)$ holds.
   U, then, executes the COMPARE COMMITMENTS protocols with bases of order N between a and b and between b and c against trustees T.
5. U executes the VSS protocol with k threshold described in subsection 2.2 for $a = BC_g(x, r_1)$. Finally, T_i receives the share (s_i, t_i) $(i = 1, \ldots, n)$ along with E_j $(j = 1, \ldots, k - 1)$.
6. Any k shareholds can recover x or factor N. (Since N is not prime, if a value in the calculating process of recovering x is not coprime to N, x cannot be recovered. But, in this case, N can be factored.) If x is recovered, N can be also factored by calculating $\gcd(x^2 \bmod N - y, N)$.

Theorem 3.1 *Let U follow the protocol. Then the above-mentioned protocol (except shares (s_i, t_i) $(i = 1, \ldots, n)$) agaist any trustees T^* is perfectly witness-indistinguishable regarding (x, r_1) with $a = BC_g(x, r_1)$.*

Theorem 3.2 *Assume that the discrete logarithm problem is intractable for any polynomial time bounded user U^*.*
Let k of n trustees T_i $(i = 1, \ldots, n)$ follow the protocol and accept each share. Then for any user, U^*, the k trustees can recover the secret key, (P, Q), that corresponds to public key $N = PQ$ with overwhelming probability.

Note: Even if the order of g is P (or Q), k trustees can recover (P, Q), since $x^4 \equiv y^2 \pmod P$. If $x^4 \equiv y^2 \pmod N$, then the above-mentioned procedure can be used to get (P, Q). Otherwise, $(x^4 - y^2, N)$ should be P.

---

[2] The trustees can jointly generate the parameters. In this paper, however, we do not care about the way how to jointly generate them.

4 Scheme 2 for RSA

This section shows another threshold key recovery protocol, Scheme 2, for the RSA scheme. Before describing this protocol, we will introduce another type of the bit-commitment protocols.

4.1 Bit-Commitment Protocols

The protocols in this section show that two bit-commitments with different modulus values are committed to the same value, while the previous protocols assume that the bit-commitments share the same modulus value.

First, suppose two bit-commitments with different modulus values: one is the same as the previous ones, $BC_g(s,r)$, with prime p as modulus, and the other is

$$BC_G'(s,r) = G^s H^r \bmod N,$$

where N is a composite (or modulus of the RSA scheme, i.e., $N = PQ$ and P, Q are primes), and $G, H \in Z_N$ and $r \in Z_{\phi(N)}$ ($\phi(N) = (P-1)(Q-1)$). Here we suppose that the committer of $BC_G(s,r)$ knows P and Q.

Next, we introduce a protocol in which prover U can prove to verifier T in a zero-knowledge manner that s_1 with $BC_g'(s_1, r_1)$ and s_2 with $BC_G'(s_2, r_2)$ are equivalent.

Let the interval be $I = [I_1, I_2]$ ($= \{i | I_1 \leq i \leq I_2\}$), $E = I_2 - I_1$, and $I \pm E = [I_1 - E, I_2 + E]$.

Protocol: COMPARE COMMITMENTS (Unshared Modulus)

Common input: x, x' and (p, q, g, h, N, G, H, I).

What to prove: U knows (s, r, r') such that $x = BC_g'(s, r)$, $x' = BC_G'(s, r')$ and $s \in I \pm E$.

Execute the following t times:

1. U chooses v_1 uniformly in $[0, E]$, and sets $v_2 = v_1 - E$. U sends to T the unordered pair of commitments $(V_1, V_1'), (V_2, V_2')$, where each component of the pair is ordered and $(V_i, V_i') = (BC_g(v_i, w_i), BC_G(v_i, w_i'))$. Here, $w_i \in_U Z_q$ and $w_i' \in_U Z_{\phi(N)}$.
2. T selects a bit $\beta \in_U \{0, 1\}$ and sends it to U.
3. U sends to T one of the following:

   (a) if β is 0, opening of both (V_1, V_1') and (V_2, V_2')
   (b) if β is 1, opening of $x \cdot V_i \bmod p$ and $x' \cdot V_i' \bmod N$ ($i \in \{1, 2\}$), such that $s + v_i \in I$. That is, $(s + v_i)$, $(r + w_i \bmod q)$ and $(r' + w_i' \bmod \phi(N))$ are revealed.
4. T checks the correctness of U's messages.

Note: When N is output by the key generation algorithm of the RSA scheme, the distributions of $w_i' \in_U Z_{\phi(N)}$ and $z \in_U Z_N$ are statistically indistinguishable.

4.2 Scheme 2 protocol

We suppose the RSA scheme as public-key cryptosystem. User U generates a pair of secret and public keys of the RSA scheme, (P, Q, d) and (N, e), where $N = PQ$ and $ed \equiv 1 \pmod{\phi(N)}$.

1. U generates a pair of secret and public keys of the RSA scheme, (P, Q, d) and (N, e). Here, U sets d such that $N \le d < 2N$. U sends (N, e) to the trustees.
2. The trustees generate (p, q, g, h) such that p is a prime, $p - 1$ is divisible by prime q, and $\mathrm{Ord}(g) = \mathrm{Ord}(h) = q$. Here, $q > 3N$. They send (p, q, g, h) to U. Here, $BC_g(s, r) = g^s h^r \bmod p$.
   The trustees randomly generate $(X_i, H) \in_U Z_N$ $(i = 1, \ldots, t)$, calculate $G_i = X_i^e \bmod N$, and send (X_i, H) to U $(i = 1, \ldots, t)$. Here, $BC_{G_i}(s, r) = G_i^s H^r \bmod N$ and $t = O(|N|)$.
3. U generates a random number $r_1 \in_U Z_q$ and calculates a and sets b_i as follows:

$$a = BC_g(d, r_1),$$

$$b_i = BC_{G_i}(d, 0) = X_i.$$

U executes the compare-commitment protocol (unshared modulus) between a and b_i $(i = 1, \ldots, t)$ with interval $I = [N, 2N]$ against trustees. Here, in the i-th round of the protocol, b_i is used $(i = 1, \ldots, t)$.
U also executes VSS protocol for a and T_i receives the share (s_i, t_i) $(i = 1, \ldots, n)$.

Theorem 4.1 *Let U follow the protocol. Then for any trustee, T_i^*, there exists a probabilistic poly time Turing machine (simulator), M with using T_i^* as a black-box, such that $M^{T_i^*}(N, s_i, t_i)$ is statistically indistinguishable from $View_{T_i^*}(N, s_i, t_i)$.*

Theorem 4.2 *Assume that the discrete logarithm problem is intractable for any polynomial time bounded user U^*. The probability is taken over the randomness of U^*, and the selection of (X_i, H) $(i = 1, \ldots, t)$.*
Let k of n trustees T_i $(i = 1, \ldots, n)$ follow the protocol and accept each share. Then for any user, U^*, the k trustees can recover the secret key. d, that corresponds to public key (N, e) with overwhelming probablity.

4.3 Scheme 2 (computational zero-knowledge version)

1. U generates a pair of secret and public keys of the RSA scheme, (P, Q, d) and (N, e). Here, U sets d such that $N \le d < 2N$. U sends (N, e) to the trustees.
2. The trustees generate (p, q, g) such that p is a prime, $p - 1$ is divisible by prime q, and $\mathrm{Ord}(g) = q$. Here, $q > 3N$. They send (p, q, g) to U.
   They also generate $X_i \in_U Z_N$ $(i = 1, \ldots, t)$, calculate $G_i = X_i^e \bmod N$, and send X_i to U $(i = 1, \ldots, t)$.
3. U calculates a, b as follows:

$$a = g^d \bmod p,$$

$$b_i = G_i^d \bmod N = X_i.$$

U executes the computational zero-knowledge version of the compare-commitment protocol (unshared modulus) between a and b_i $(i = 1, \ldots, t)$ with interval $I = [N, 2N]$ against trustees [3].
U also executes the VSS protocol (conditionally secure version [11]) for a.

5 Scheme 3 for RSA and Rabin

The following values are determined and published by the system.

p, q: primes, $q|p - 1$, $\text{Ord}(g) = \text{Ord}(h) = q$, $BC_g(x, r) = g^x h^r \bmod p$

We suppose the RSA and Rabin schemes as public-key cryptosystem. User U generates a pair of secret and public keys of the RSA or Rabin scheme, (P, Q) and $N = PQ$. We assume that $(-1/N) = 1$. We also assume that $|q| > |N|$ (see [5] for the detailed relationships between $|q|$ and $|N|$).

1. U generates two primes P and Q, and calculates $N = PQ$. U sends N to the trustees.
2. U also gives the evidence to the trustees that N consists of two primes [8, 2].
3. U calculates a, b, c as follows:

$$a = BC_g(P, r_1),$$

$$b = BC_g(Q, r_2),$$

$$c = BC_g(N, r_3) = BC_a(Q, r_3 - r_1 Q)$$

U releases (a, b, r_3) to the trustees.

U, then, executes the check-commitment protocol (Appendix) for a and the compare-commitment protocol (Appendix) between b and c with trustees.

4.
$$x = BC_g(\alpha, r_4),$$

$$y = BC_g(\beta, r_5),$$

$$z = BC_a(\alpha, r_6) = BC_g(\alpha P, r_6'),$$

$$w = BC_b(\beta, r_7) = BC_g(\beta Q, r_7'),$$

U opens $zw \bmod p = BC_g(1, r_6' + r_7')$.

U, then, executes the compare-commitment protocols (Appendix) between x and z and between y and w with trustees.

5. U executes the VSS protocol for a, and T_i receives the share (s_i, t_i) ($i = 1, \ldots, n$).

Theorem 5.1 *Let U follow the protocol. Then for any trustee, T_i^*, there exists a probabilistic poly time Turing machine (simulator), M, with using T_i^* as a black-box, such that $M^{T_i^*}(N, s_i, t_i)$ is perfectly indistinguishable from $View_{T_i^*}(N, s_i, t_i)$.*

Theorem 5.2 *Assume that the discrete logarithm problem is intractable for any polynomial time bounded user U^*.*

Let k of n trustees T_i ($i = 1, \ldots, n$) follow the protocol and accept each share. Then for any user, U^*, the k trustees can recover the secret key, (P, Q), that corresponds to public key $N = PQ$ with overwhelming probability.

6 Evaluations

In this section, we roughly evaluate the efficiency of the proposed three schemes. Scheme 1 requires:

1. (Evidence for N with two primes: Non-interactive version)
   If 50 random numbers with positive Jacob symbol values are examined, then in average 25 modular square roots are required to compute. (Note that the Jacob symbol test is much faster than the square rooting.)
   $25 \times |N|$ should be sent to trustees and stored.
2. (Main part)
   Two compare-commitment protocols are required to execute. In each protocol, 2 modular exponentiations (with tow or three bases) are required (for U or T). Hence, totally 7 $(= 2 \times 2 + 3)$ modular exponentiations are required. The required communication amount in each protocol is $(2|p| + 3|N|)$ bits. Hence, totally $(7|p| + 6|N|)$ bits are required for communication.
3. (VSS part)
   $k - 1$ modular exponentiations (with tow bases) are required for U.
   Each trustee stores $|N|$ and $|p|$.

Scheme 2 requires:

1. (Main part)
   One compare-commitment protocols are required to execute. Hence, totally $4t + 2$ modular exponentiations are required.
   As for communication complexity, $(2t|p| + 4t|q| + 6t|N| + |p| + |N|)$ bits are required for communication.
2. (VSS part)
   Same as Scheme 1.

Scheme 3 requires:

1. (Evidence for N with two primes: Non-interactive version)
   Same as Scheme 1.
2. (Main part)
   Three compare-commitment protocols and one check-commitment protocol are required to execute. In a compare-commitment protocol, $4t$ modular exponentiations (with tow bases) are required (for U or T). In a a check-commitment protocol, $2t$ modular exponentiations (with tow bases) are required (for U or T). Hence, totally $14t + 7$ modular exponentiations are required.
   As for communication complexity, $(14t|p| + 28t|q| + 7|p|)$ bits are required for communication.
3. (VSS part)
   Same as Scheme 1.

7 Conclusion

This paper proposed the first practical threshold key-recovery systems for the RSA and Rabin schemes. Although these schemes are less efficient than the threshold key-recovery system for the ElGamal scheme, they are still very practical.

References

1. Blakley, G.R.: Safeguarding Cryptographic Keys, Proc. of AFIPS 1979 Nat. Computer Conf., vol.48, pp.313–317 (Sep. 1979)
2. Blum, M. : Coin Flipping by Telephon, Proc. of COMPCON, IEEE, pp.133-137 (1982).
3. Brickell, E., Chaum, D., Damgård, I. and van de Graaf, Gradual and Verifiable Release of a Secret, Proc. of Crypto 87, LNCS, Springer-Verlag (1988).
4. Chor, B., Goldwasser, S., Micali, S. and Awerbuch, B.: Verifiable Secret Sharing and Achieving Simultaneity in the Presence of Faults, Proc. of FOCS, pp.383-395 (1985).
5. Damgård, I.: Practical and Provably Secure Release of a Secret and Exchange of Signatures, Proc. of Eurocrypt'93, LNCS 765, Springer-Verlag, pp.200-217 (1994).
6. Feldman, P.: A Practical Scheme for Non-Interactive Verifiable Secret Sharing, Proc. of FOCS'87, pp.427-437 (1987).
7. Fiat, A. and Shamir, A.: How to Prove Yourself: Practical Solutions to Identification and Signature Problems, Proc. of Crypto'86, LNCS 263, Springer-Verlag, pp.186-194 (1986).
8. Micali, S.: Fair Public-Key Cryptosystems, Proc. of Crypto'92, LNCS, Springer-Verlag, pp.113-138 (1993).
9. Okamoto, T.: Provably Secure and Practical Identification Schemes and Corresponding Signature Schemes, Proc. of Crypto'92, LNCS 740, Springer-Verlag, pp.31-53 (1993).
10. Okamoto, T.: An Efficient Divisible Electronic Cash Scheme, Proc. of Crypto'95, LNCS 963, Springer-Verlag, pp.438-451 (1995).
11. Pedersen, T. P.: Distributed Provers with Applications to Undeniable Sigantures, Proc. of Eurocrypt'91, LNCS 547, Springer-Verlag, pp.221-242 (1991).
12. Pedersen, T. P.: Non-Interactive and Information-Theoretic Secure Verifiable Secret Sharing, Proc. of Crypto'91, LNCS 576, Springer-Verlag, pp.129-140 (1992).
13. Shamir, A.: How to Share a Secret, Comm. Assoc. Comput. Mach., vol.22, no.11, pp.612–613 (Nov. 1979)

Appendix: Conventional Bit-Commitment Protocols

Here we introduces some conventional bit-commiment protocols in which U can prove to T in a zero-knowledge manner that a committed value is in an interval, and that two committed values are equivalent.

Let the interval be $I = [a, b]$ $(= \{i | a \leq i \leq b\})$, $e = b - a$, and $I \pm e = [a - e, b + e]$.

Protocol: CHECK COMMITMENT
Common input: x and (p, q, g, h, I).
What to prove: U knows (s, r) such that $x = BC_g(s, r)$ and $s \in I \pm e$.
Execute the following t times:

1. U chooses v_1 uniformly in $[0, e]$, and sets $v_2 = v_1 - e$. U sends to B the unordered pair of commitments $V_1 = BC_g(v_1, w_1)$, $V_2 = BC_g(v_2, w_2)$. (Let X = unordered pair of V_1 and V_2.)
2. T selects a bit $\beta \in_U \{0, 1\}$ and sends it to U.
3. U sends to T one of the following (say, Y):

   (a) if β is 0, opening of both V_1 and V_2
   (b) if β is 1, opening of $x \cdot V_i \bmod p$ $(i = 1, 2)$, such that $s + v_i \in I$.

4. T checks the correctness of U's messages.

Protocol: COMPARE COMMITMENTS
Common input: x, x' and (p, q, g, a, h, I), where the order of a is q.

What to prove: U knows (s, r, r') such that $x = BC_g(s, r)$, $x' = BC_a(s, r')$ and $s \in I \pm e$.

Execute the following t times:

1. U chooses v_1 uniformly in $[0, e]$, and sets $v_2 = v_1 - e$. U sends to T the unordered pair of commitments (V_1, V_1'), (V_2, V_2'), where each component of the pair is ordered and $(V_i, V_i') = (BC_g(v_i, w_i), BC_a(v_i, w_i'))$. (Let $X =$ unordered pair of (V_1, V_1') and (V_2, V_2').)
2. T selects a bit $\beta \in_U \{0, 1\}$ and sends it to U.
3. U sends to T one of the following (say, Y):

   (a) if β is 0, opening of both (V_1, V_1') and (V_2, V_2')
   (b) if β is 1, opening of $x \cdot V_i \bmod p$ and $x' \cdot V_i' \bmod p$ $(i = 1, 2)$, such that $s + v_i \in I$.

4. T checks the correctness of U's messages.

Non-interactive version Non-interactive versions of CHECK COMMITMENT and COMPARE COMMITMENTS protocols employ the standard conversion technique using hash functions. Let h be an arbitrary hash function.

First, U generates X repeatedly t times (say, $X_1, \ldots, X_t$). Then, set $(\beta_1, \ldots, \beta_t) = h(X_1, \ldots, X_t)$. Finally, Y_i is calculated from X_i and β_i $(i = 1, \ldots, t)$, and $(Y_1, \ldots, Y_t)$ is obtained. Then, the non-interactive proof for CHECK COMMITMENT or COMPARE COMMITMENTS protocol is $(X_1, \ldots, X_t; Y_1, \ldots, Y_t)$.

A Weakness of the Menezes-Vanstone Cryptosystem

Klaus Kiefer

Universität des Saarlandes, Graduiertenkolleg Informatik*
Postfach 151150, D-66041 Saarbrücken (Germany)
e-mail: kiefer@cs.uni-sb.de

Abstract. In this paper we show, that the elliptic curve cryptosystem by Menezes and Vanstone is *not* really a probabilistic cipher, in contrast to its design. Each ciphertext leaks some kind of information, which could be used for unauthorized decryption, if the cryptosystem is set up in a careless way. But in any case we have a loss of efficiency, since the additional effort, which always comes with probabilistic encryption, does not pay.

1 Introduction

Elliptic curves are popular settings for building efficient public key cryptosystems, since in general computing discrete logarithms in these groups is difficult. If we have, for example, a non-supersingular curve over a prime field $\mathbf{F}_p$, for which the discrete logarithm problem is hard (this can be checked efficiently [6]), DL is intractable for p with about 45 (or more) decimal digits. In comparison, factorization of hard composite numbers, whose difficulty guarantees the security of the RSA cryptosystem for example, is tractable for numbers up to 130 digits. Therefore elliptic curve cryptosystems can be set up with numbers of relatively small size, a possible gain in efficiency. But on the other hand: if messages are encoded as curve points (as in the ElGamal scheme), we have a loss of efficiency, since by the well known Hasse-Theorem an elliptic curve has approximately as many elements as the underlying field, but we need two field elements for representation of each point. Further no deterministic algorithm is known for computing such an encoding of plaintexts - there are probabilistic algorithms, which use only a small subset of all points [3]. To get rid of these disadvantages, as a variation of ElGamal the Menezes-Vanstone cryptosystem was proposed. Here curve points serve as masks

---

* Author is member of research group of Prof. J. Buchmann, who moved to Technische Hochschule Darmstadt. Graduiertenkolleg Informatik is granted by Deutsche Forschungsgemeinschaft (DFG).

for message blocks, which are arbitrary pairs of numbers from the used field. In this paper we will show that there are some problems with this approach.

2 Technical Background

Definition 2.1 ((DL Problem)). Let G be a group (multiplicatively written) and $\alpha \in G$. By the Discrete Logarithm Problem we denote the task of finding for any $\beta \in \langle \alpha \rangle = \{\alpha^0, \alpha^1, \alpha^2, \ldots\}$ the minimal exponent $k \geq 0$ with $\alpha^k = \beta$.

The security of the ElGamal cryptosystem [2] seems closely related to the difficulty of the DL problem, which is believed to be computationally infeasible for some suitable groups.

Definition 2.2 ((General ElGamal Cryptosystem)). Let G be a group and α be an element from G, such that the corresponding DL problem is intractable. Then the ElGamal Cryptosystem is set up as follows:

- set of plaintexts: $P = G$.
- set of ciphertexts: $C = \langle \alpha \rangle \times G$.
- set of keys: $K = \{(a, \beta) \mid \beta = \alpha^a\}$.
  a is the private key, whereas β is public.
- encryption procedure: to encrypt $x \in P$ under the key $k \in K$, Alice first has to choose a (secret) random number $r \in \{1, 2, \ldots, \mathrm{ord}(\alpha) - 1\}$. Then: $e_k(x, r) = (\alpha^r, x\beta^r)$.
- decryption procedure: Bob, who knows the private key a, computes the plaintext by: $d_k(y_0, y_1) = y_1 (y_0^a)^{-1}$.

There are two typical settings for the ElGamal scheme:

1. $G = (\mathbf{F}_q^*, \cdot)$, where $\mathbf{F}_q$ is the field with $q = p^n$ elements. By the Index Calculus Method (see [5] or [7]) we have a subexponential algorithm for solving the DL problem in these groups, which has been improved by many variations.
   It is easy to see that breaking the ElGamal scheme is equivalent to the well known Diffie-Hellman problem [10]. In the case of a prime field ($n = 1$) it was proved that the latter is as hard as the DL problem for "many" p [4], so we *know* that breaking ElGamal is not easier than DL for that class of prime numbers. The record for computing a hard DL problem (i.e. not suitable for the Pohlig-Hellman algorithm [8])

in the multiplicative group of a prime field was set for a p with 85 decimal digits [11]. If we have a non-prime field, then there are faster algorithms for special cases, e.g. [1].

2. $G = (E, +)$, where E is an elliptic curve over $\mathbf{F}_q$. As there are no better than exponential algorithms in this case (see [7] or [8]), the corresponding DL problem is intractable for q having more than, say, 45 digits. From this point of view the ElGamal scheme can be set up with numbers of relative small size, a possible gain in efficiency. On the other hand now we have a message expansion of about four - compared to factor two in the ElGamal scheme over $\mathbf{F}_q$ - since by the well known Hasse-Theorem $(\mid \#E - (q+1) \mid \leq 2\sqrt{q})$ there are $\#E \approx q$ possible plaintexts, whereas each ciphertext consists of four numbers from $\mathbf{F}_q$ [10]. Moreover no deterministic algorithm is known for computing an encoding of plaintexts as points of an elliptic curve. There are probabilistic algorithms, which hit (for a sufficiently small failing probability) only a small fraction of all points [3], a further loss in efficiency. To get rid of these disadvantages, the following cryptosystem was proposed (see [10]).

Definition 2.3 ((Menezes-Vanstone Cryptosystem)). Let E be an elliptic curve over $\mathbf{F}_q$ and $\alpha \in E$, such that the corresponding DL problem is intractable. Then the Menezes-Vanstone Cryptosystem is set up as follows:

- set of plaintexts: $P = \mathbf{F}_q^* \times \mathbf{F}_q^*$.
- set of ciphertexts: $C = \langle \alpha \rangle \times (\mathbf{F}_q^* \times \mathbf{F}_q^*)$.
- set of keys: $K = \{(a, \beta) \mid \beta = a\alpha\}$.
  a is the private key, whereas β is public.
- encryption procedure: to encrypt $x \in P$ under the key $k \in K$, Alice first has to choose a (secret) random number $r \in \{1, 2, \ldots, \operatorname{ord}(\alpha) - 1\}$ (restriction: $r\beta$ must have non-zero components, i.e. $r\beta \in \mathbf{F}_q^* \times \mathbf{F}_q^*$). Then she computes $e_k(x, r) = (r\alpha, x \otimes (r\beta))$, where $\otimes$ denotes coordinate-wise multiplication in $\mathbf{F}_q^* \times \mathbf{F}_q^*$.
- decryption procedure: Bob, who knows the private key a, computes the plaintext by: $d_k(y_0, y_1) = y_1 \otimes (ay_0)^{-1}$ (inversion in $\mathbf{F}_q^* \times \mathbf{F}_q^*$ is also done coordinate-wise).

In this cryptosystem the problem of embedding plaintexts as points of an elliptic curve is removed, since points serve only as masks for plaintext blocks (which are just pairs of numbers from $\mathbf{F}_q^*$ - thereby message expansion is reduced to factor two and efficient encoding of plaintexts is easy), and masking is done by another than the group operation. In the next section we show that there are some problems with this approach.

3 Security Issues

In the Menezes-Vanstone cryptosystem, which uses randomized encryption, a guessed decryption of a ciphertext can be easily checked by anyone. This is also true for every deterministic, but not for probabilistic public key systems like ElGamal. This "decryption-checking" test is probabilistic and no-biased, i.e. an answer "no" (wrong plaintext) is always true, whereas "yes" is correct with high probability. It works as follows:

> dc_test
>
> INPUT: elliptic curve E over $\mathbf{F}_q$,
>
> ciphertext $y = (y_0, y_1)$, $y_1 = (z_1, z_2)$,
>
> guessed plaintext $x = (x_1, x_2)$
>
> OUTPUT: "yes", if $c := y_1 \otimes x^{-1} = (z_1 x_1^{-1}, z_2 x_2^{-1}) \in E$
>
> "no", else

It is easy to see that the answer "no" is always correct. If x is not the correct decryption, c is any pair from $\mathbf{F}_q^* \times \mathbf{F}_q^*$, and by the Hasse-Theorem the chance for an answer "yes" is given by:

$$\Pr_{c \in \mathbf{F}_q^* \times \mathbf{F}_q^*} (c \in E) \leq \frac{q + 1 + 2\sqrt{q}}{(q-1)^2} = \mathcal{O}(1/q)$$

So we have a very small error probability.

With the above test we have not broken the Menezes-Vanstone cryptosystem in general. But one has to be careful: probabilistic encryption can be used, if the message space is small. In this case the dc_test allows decryption by exhaustive search.

A similar situation might occur, if the cryptosystem is set up with numbers of small size while encrypting natural language. We need the following:

Definition 3.1 ((Entropy)). Let X be a random variable in n values $x_1, \ldots, x_n$. Then the entropy $h(x)$ of X is defined by

$$h(X) = -\sum_{i=1}^{n} p(x_i) \log_2(p(x_i)).$$

The entropy of a natural language L measures the information (in bits) per letter and is given by

$$h(L) = \lim_{n \to \infty} \frac{h(T_n)}{n},$$

where T_n is the random variable for text fragments consisting of n letters. If we set up the cryptosystem over $\mathbf{F}_q$ with q having s decimal digits, each plaintext block consists of $2s \log_2 10 \approx 6.644s$ bits. If ASCII code is used (8 bits per letter), then there are $0.83s$ letters per block, and with Shannon's estimate ($H(\text{English}) = 1.2$ [9]) we find that there are about 2^s possibilities for each block. For $s \approx 45$ this amount might be testable by exhaustive search.

In any case we see that each ciphertext leaks some kind of information about the corresponding plaintext. But in a probabilistic cryptosystem the possible encryptions of two different plaintexts should "look like" each other. More formally: they must not be ϵ-distinguishable [10] for any noticeable ϵ.

Definition 3.2 ((ϵ-distinguishability)). Let Y be a finite set, p_0 and p_1 be probability distributions on Y and $A : Y \to \{0, 1\}$ be a (probabilistic) algorithm. The expected average answers are given by:

$$E_A(p_j) := \sum_{y \in Y} p_j(y) \cdot E(A[y]), \quad j = 0, 1,$$

where $E(A[y])$ denotes the expected output of A, given input y. A is called a ϵ-distinguisher of p_0 and p_1, if

$$|E_A(p_0) - E_A(p_1)| \geq \epsilon.$$

p_0 and p_1 are called ϵ-distinguishable, if there is an ϵ-distinguisher of them.

Theorem 3.3. *The Menezes-Vanstone scheme is not a probabilistic cryptosystem.*

Proof. Let $p_{k,x}(y)$ denote the probability that y is an encryption of x under the key k. For a probabilistic cryptosystem it must hold, that for all $k \in K$ and for all $x \neq x' \in P$ the probability distributions $p_{k,x}$ and $p_{k,x'}$ are not ϵ-distinguishable for a small ϵ. This is not the case for the Menezes-Vanstone cryptosystem. Let A execute the dc_test with x as supposed plaintext ("yes" $= 1$, "no" $= 0$). Then $E_A(p_{k,x}) = 1$ (as the

test is no-biased, for a correct decryption the answer is always "yes")
and $E_A(p_{k,x'}) \approx 1/q$ (error probability of dc_test; note that possible encryptions of any plaintext are equally distributed, since choosing different
random values leads to different ciphertexts). Therefore $p_{k,x}$ and $p_{k,x'}$ are
ϵ-distinguishable for $\epsilon \approx 1 - 1/q$.

4 Conclusion

We have shown a weakness of the Menezes-Vanstone cryptosystem, which
might make this cipher insecure under certain circumstances. In any case
it is not a probabilistic cryptosystem, the additional effort for randomized
encryption does not pay. And we saw that a basic problem in using
elliptic curve cryptosystems - coding plaintexts as curve points - cannot
be removed in such an easy way.
The author is grateful to V. Müller and D. Weber for helpful comments.

References

1. D. Coppersmith. *Fast Evaluation of Discrete Logarithms in Fields of Characteristic Two*, IEEE Transactions of Information Theory IT-30 (1984), pp. 587-594.
2. T. ElGamal. *A Public Key Cryptosystem and a Signature Scheme Based on Discrete Logarithms*, IEEE Transactions on Information Theory, **31** (1985), pp. 469-472.
3. N. Koblitz. *A Course in Number Theory and Cryptography*, New York, 1987.
4. U. Maurer, S. Wolf. *Diffie-Hellman-Oracles*, Advances in Cryptology - CRYPTO'96 Proceedings, pp. 268-282.
5. K. McCurley. *The Discrete Logarithm Problem*, Cryptology and Computational Number Theory, AMS Proc. Symp. in Applied Mathematics, **42** (1990), pp. 49-74.
6. A. Menezes, S. A. Vanstone. *Elliptic Curve Cryptosystems and Their Implementation*, Journal of Cryptology, **6** (1993), pp. 209-224.
7. A. Odlyzko. *Discrete Logarithms and their Cryptographic Significance*, Advances in Cryptology - EUROCRYPT'84 Proceedings, pp. 224-314.
8. S. Pohlig, M. Hellman. *An Improved Algorithm for computing Logarithms over $GF(p^n)$ and its Cryptographic Significance*, IEEE Transactions of Information Theory, **24** (1978), pp. 106-110.
9. C. E. Shannon. *Prediction and Entropy in Printed English*, Bell System Technical Journal, **30** (1951), pp. 50-64.
10. D. R. Stinson. *Cryptography: Theory and Practice*, Boca Raton, 1995.
11. D. Weber, T. Denny, J. Zayer. *Discrete Log Record*, posting to Number Theory Net (NMBRTHRY@listserv.nodak.edu), November 25, 1996.

On Ideal Non-perfect Secret Sharing Schemes

Pascal Paillier

Gemplus - Cryptography Department
1 Place de la Méditerrannée
BP 636, F-95206 Sarcelles cedex- France
106333,7300compuserve.com

Abstract. This paper first extends the result of Blakley and Kabatian-ski [3] to general non-perfect SSS using information-theoretic arguments. Furthermore, we refine Okada and Kurosawa's lower bound [12] into a more precise information-theoretic characterization of non-perfect secret sharing idealness. We establish that in the light of this generalization, ideal schemes do not always have a matroidal morphology. As an illustration of this result, we design an *ad-hoc* ideal non-perfect scheme and analyze it in the last section.

1 Introduction

A secret sharing scheme [13], [14] is a protocol by the means of which n share-holders $P = \{1, \cdots, n\}$ share a secret information s. Historically, the first secret sharing algorithms solved the following problem :

Problem 1. Split s into n shares in such a way that

- the knowledge of at least k shares makes s computable,
- the knowledge of $k - 1$ shares leaks no information about s.

The first protocols featuring these properties (also known as (k, n)-*threshold secret sharing schemes*) were exhibited by Shamir [13] and Blakley [2] in the late seventies and received considerable attention because of their clear mathematical formulation and potential military and civilian applications [6, 7, 8]. A first extension considers the following (more general) problem :

Problem 2. Split s into n shares in such a way that

- participants belonging to some *authorized* subsets of P can join forces and recover s,
- all other coalitions can not infer any information about s.

The set of authorized coalitions, denoted $\Gamma \subset 2^P$ and called *access structure*, is a basic scheme-parameter. Protocols that comply with the requirements of problem 2 are *perfect* when protection against illegitimate recovery of s is guaranteed by unconditional information-theoretic considerations. Karnin *et al.* [10] showed that in all perfect schemes and for all access structures, shares must be bigger than s. Consequently, *ideal* perfect schemes were naturally defined as those optimal cases where the shares and s are of equal size. Deciding if, given an arbitrary access structure Γ, an ideal solution of problem 2 exists, is still an open problem.

Traditionally, non-perfect schemes are those protocols where illegitimate coalitions are neutralized by computational (rather than unconditional) infeasibility :

Problem 3. Split s into n shares such that

- participants belonging to $A \in \Gamma$ can join forces and recover s,

- other coalitions can not compute s with reasonable computational resources.

The main interest of non-perfectness lies in the ability to bound the share-size at the cost of a theoretically weakened but practically acceptable security[1]. This particular feature provides an efficient tool against *greedy* access structures for which shares have to be much bigger than the secret [5].

For the sake of further refinement, several authors bounded the conditions under which non-perfectness and minimal size co-exist. In this respect, Kurosawa *et al.* [11] defined K-ideal non-perfect schemes on strict informational grounds and exhibited a direct link between ideal perfect and K-ideal non-perfect schemes.

The purpose of this paper is to shed a different light on the relationship between non-perfect idealness and secret sharing matroids. We will first extend [3]'s main theorem to non-perfect schemes, by prooving that the distribution rule of secrets plays no role in the idealness-matroid correspondence. Furthermore, we will show that refining K-idealness into a much more accurate idealness conception unfortunately destroys this correspondence. This suggests inescapable limitations of using matroids as analytic tools to describe the combinatorial morphology of general secret sharing schemes.

2 Previous Contributions

2.1 Perfect Sharing Schemes

More formally, given an access structure Γ on P and a set of secrets S, a perfect secret sharing scheme $\mathcal{S}(P, \Gamma, S)$ is a pair $\{D, R\}$ of probabilistic algorithms. The

---

[1] it is worthwhile noting the existence of similar phenomena in completely different scientific disciplines : a characteristic example is lossy compression (which is to some extent, non-perfect) which trades-off better compression rates against acceptable information loss.

distribution algorithm $D(s,i) = v_i$ takes as input the secret $s \in S$ and a participant index $i \in P$, and outputs the share v_i whilst the *reconstruction* process $R(\{v_i\})$ is used by a coalition to recover s. A coalition $A \subset P$ is authorized if (and only if) $R(v_A) = s$ where $v_A = \{v_i \mid i \in A\}$.

Consequently, $H(S|V_A) = 0$ where S and V_A denote the random variables induced by s and v_A and H denotes entropy[2]. In perfect schemes, $B \notin \Gamma$ implies $H(S|V_B) = H(S)$ which simply express B's absolute ignorance of s.

Recently, Itoh *et al.* [9], and Benaloh *et al.* [1] established that perfect SSS exist if and only if Γ is monotone, that is

$$A \in \Gamma \quad \text{and} \quad A \subset B \quad \Rightarrow \quad B \in \Gamma .$$

In such a case, Γ is completely characterized by the family of its minimal elements for set inclusion Γ^-. A share-holder $i \in P$ is *independent* in Γ if $i \in A$ for some $A \in \Gamma^-$. If i is dependent, his share is clearly useless and can be replaced by zero (without loss of generality, all participants are henceforth assumed independent). According to [10], one must have

$$H(V_i) \geq H(S) \tag{1}$$

for all perfect schemes. Ideal[3] perfect schemes are therefore naturally characterized by $H(V_i) = H(S)$ for all i. Finally, it is also known [4] that ideal perfect schemes have a matroidal morphology.

2.2 Non-perfect Secret Sharing Schemes

By straightforward negation, $|V_i| \leq |S|$ implies non-perfectness, in which case illegitimate coalitions may still have some partial access to the secret, *i.e.*

$$0 < H(S|V_B) < H(S) \quad \text{for some } B \subset P .$$

In other words, although B can not recover the whole secret, something still leaks out. As a further generalization, a non-perfect scheme has a level-d access hierarchy if there exists a partition of 2^P, denoted $\Sigma = (\Sigma_0, \Sigma_1, \cdots, \Sigma_d)$, such that

$$A \in \Sigma_k \quad \Leftrightarrow \quad H(S|A) = (k/d)H(S) .$$

Trivially, perfect schemes can be looked upon as non-perfect schemes which d equals one. Just as for perfect schemes, it is established that non-perfect schemes,

---

[2] to simplify notations, random variables and their domains, respectively X and $\{x \mid P(X = x) > 0\}$, will both be represented by X; consequently, $|X| = \log_2 \sharp X$ (where $\sharp X$ denotes the cardinality of X) and $|X| = H(X)$ if X is uniformly distributed.

[3] *stricto sensu*, a scheme is C-ideal (respectively IT-ideal) if $|V_i| = |S|$ (respectively $H(V_i) = H(S)$); these notions coincide when random variables are uniformly distributed ; under this assumption, [11] showed that IT-ideal and C-ideal perfect schemes are matroids ; more recently, [3] strengthened this result by proving that IT-ideal perfect schemes are matroids even when the distribution of secrets is not uniform.

denoted $\mathcal{S}(P, \Sigma, S)$, exist if and only if $\cup_{k \leq l} \Sigma_k$ is monotone for all $l \leq d$. In non-perfect schemes, i is *independent* in Σ if $\bar{i} \in A$ for some $A \in \Sigma_k^-$ (again, w.l.o.g we adopt a general independence assumption). In [11], Kurosawa *et al.* proved that

$$H(V_i) \geq H(S)/d \tag{2}$$

for all i in level-d non-perfect schemes. Naturally, K-ideal non-perfect schemes are those cases in which the previous bound is reached [4] :

$$H(V_i) = H(S)/d . \tag{3}$$

More importantly, assuming that all variables are uniformly distributed, Kurosawa *et al.* proved that K-ideal non-perfect schemes inherit a matroidal morphology from their perfect ancestors. This result, reflected a fundamental correspondence between matroidal structure and small-size shares, and the in-dependency of this relationship towards perfect or non-perfect characteristics of schemes.

Surprisingly, formulae (1) and (2) appear to be particular cases of a much more precise information theoretic framework, in the light of which, ideal schemes are matroids only under very particular circumstances.

2.3 Secret Sharing Schemes and Matroids

A *polymatroid* $T = \{Q, \phi\}$ is a system composed of a finite set Q and a *rank function* $\phi : 2^Q \longrightarrow \mathbb{R}^+$ such that for all $X, Y \subset Q$:

(2.3.1) $\phi(\varnothing) = 0$

(2.3.2) If $X \subset Y$ then $\phi(X) \leq \phi(Y)$

(2.3.3) $\phi(X \cup Y) + \phi(X \cap Y) \leq \phi(X) + \phi(Y)$

matroids are those polymatroids which rank function also verifies :

(2.3.4) $\phi(X) \in \mathbb{N}$

(2.3.5) If $\sharp X = 1$ then $\phi(X) \leq 1$

Alternatively, a matroid on Q can also be seen as a collection Ω of subsets of Q such that

(2.3.6) $\varnothing \in \Omega$

(2.3.7) If $A \in \Omega$ and $B \subset A$ then $B \in \Omega$

(2.3.8) If $A, B \in \Omega$ and $\sharp A = \sharp B + 1$ then $\exists j \in A \setminus B$ s.t. $j \cup B \in \Omega$

---

[4] as for perfect schemes, one can naturally envision non-perfect CK-idealness ($|V_i| = |S|/d$) and non-perfect ITK-idealness ($H(V_i) = H(S)/d$), but [11] seems to focus on the sole setting in which both notions converge into K-idealness.

ϕ is then defined by

$$\phi(A) = \max\{\sharp X \mid X \subset A, X \in \Omega\}.$$

Theorem 4 [5]. *Perfect schemes have a polymatroid structure.*

Let $\mathcal{S}(P, \Gamma, S)$ be a perfect scheme, $Q = S \cup P$ and $\phi(A) = H(A)/H(S)$ for $A \subset Q$. $\{Q, \phi\}$ is then a polymatroid.

Theorem 5 [11]. *If S is uniformly distributed, ideal perfect schemes have a matroid structure.*

If the scheme is ideal, $|V_i| = H(i) = H(S) = |S|$ which implies $\phi(i) = 1$. A proof that $\phi(A) \in \mathbb{N}$ can be found in [11]. It is also worthwhile to observe that :

$(P_1) \qquad \phi(SA) - \phi(A) = \dfrac{H(SA) - H(A)}{H(S)} = \dfrac{H(S|A)}{H(S)} = \begin{cases} 0 & \text{if } A \in \Gamma \\ 1 & \text{otherwise} \end{cases}$

Theorem 6 [3]. *Ideal perfect schemes have a matroid structure.*

Theorem 7. *Non-perfect schemes have a polymatroid structure.*

Let $\mathcal{S}(P, \Sigma, S)$ be a non-perfect scheme, and $S = S_1 \| \cdots \| S_d$ where $\|$ stands for concatenation. Define $Q = \{S_1, \cdots, S_d\} \cup P$ and let[5]

$$\phi(A) = \frac{H(A)}{H(S)} \times d \quad \text{for } A \subset Q \qquad (4)$$

$\{Q, \phi\}$ is a polymatroid since ϕ verifies properties (2.3.1) to (2.3.2), and the proof arguments of theorem 4 still hold.

Theorem 8 [11]. *If S is uniformly distributed, K-ideal non-perfect schemes have a matroid structure.*

Property (2.3.5) follows from equation (3) since $\phi(i) = d \times H(i)/H(S) = 1$. ϕ's integer-valuation proof (rather heavy and technical) can be found in [11]. It is easy to observe that ϕ also verifies the following property

$(P_2) \qquad\qquad\qquad \phi(S_1 \cdots S_d A) - \phi(A) = d(A)$

which is an extension of (P_1)[6].

In the next section, we will establish that K-ideal schemes present a matroid structure in any case, without appealing to a particular distribution on secrets, *i.e*

Theorem 9. *K-ideal non-perfect schemes have a matroidal morphology.*

Since ideal perfect schemes are particular K-ideal non-perfect schemes, this result can then clearly be seen as an immediate generalization of Blackley and Kabatianski's theorem shown above as theorem 6.

---

[5]in a much more formal spirit, [11]'s authors defined the associated *mixed access hierarchy* $\widehat{\Sigma} = \{\widehat{\Sigma}_k\}_{k \leq d}$ as a natural extension of Σ on Q ; to simplify notations, we will denote Σ's extension by Σ itself.

[6]on the other hand, [11] showed that K-ideal SSS can be associated to a class of properly-chosen (representable over GF_q) matroids satisfying property (P_2).

3   K-idealness and Matroids

In this section, we generalize [3] and provide a proof that uniform secret-distribution is not a necessary correspondence condition between K-idealness and matroids. Let $\mathcal{S}(P, \Sigma, S)$ be a level-d non-perfect scheme. Denoting $d(A) = k$ when $A \in \Sigma_k$, and letting[7]

$$\delta(i) = \max_{X \subset Q} d(X) - d(X \cup i)$$

we get [12] :

$$H(V_i) \geq \frac{\delta(i)}{d} H(S) \quad \text{for all } i \in P . \tag{5}$$

Now let,

$$\delta[\![A]\!] = \sum_{i \in A} \delta(i) .$$

Recall that $\phi(A) = H(A) \times d / H(S)$ for $A \subset Q$ and define

$$G_{\mathcal{S}(P, \Sigma, S)} = \{ A \subset Q \mid \phi(A) = \delta[\![A]\!] \} .$$

Then,

Lemma 10. *If* $\mathcal{S}(P, \Sigma, S)$ *is K-ideal, then* $G_{\mathcal{S}(P, \Sigma, S)}$ *is a matroid.*

Proof. K-idealness imposes $H(V_i) = H(S)/d$. Since inequation (5) holds, we necessarily have $\delta(i) = 1$ for all $i \in P$. Hence $\delta[\![A]\!] = \sharp A$ and we need to prove that

$$G_{\mathcal{S}(P, \Sigma, S)} = \{ A \subset Q \mid \phi(A) = \sharp A \}$$

is a matroid. But,

- (2.3.1) is straightforward,

- Since $H(A) \leq \sum_{i \in A} H(i)$, we have $\phi(A) \leq \sum_{i \in A} \phi(i) = \sum_{i \in A} 1 = \sharp A$ for all $A \subset Q$. Now take $A \in G$ and $j \in A$. It comes

$$\sharp A = \phi(A) \leq \phi(A \setminus j) + \phi(j) = \phi(A \setminus j) + 1$$

which gives $\sharp A - 1 \leq \phi(A \setminus j) \leq \sharp(A \setminus j)$. Therefore $\phi(A \setminus j) = \sharp A - 1$ and $A \setminus j \in G$. Immediate induction yields $B \in G$ for all $B \subset A$: condition (2.3.2) is met.

- The proof that condition (2.3.3) is fulfilled will appear in the final version of the paper.   □

                                                                                  □

---

[7]note that $\delta(i)$ (the maximal "knowledge-lag" in $\Sigma = \{\Sigma_k\}_{k < d}$ that i may ever be held responsible for during the secret recovery process) only depends on the access hierarchy Σ.

Remark. The rank function associated to $G_{\mathcal{S}(P,\Sigma,S)}$ is ϕ. When S is uniformly distributed, $A \longmapsto \phi(A)$ defines a matroid on Q by *virtue* of theorem 8. According to lemma 10, this matroid and $G_{\mathcal{S}(P,\Sigma,S)}$ have identical rank functions. They are therefore equal. As a consequence, the previous lemma characterizes Kurosawa *et al.*'s matroidal structure discovered in [11].

Moreover, notice that in perfect SSS we have $d = 1$ which implies directly $\delta(i) = 1$, in which case $\phi(A) = H(A)/H(S)$ and $\delta[\![A]\!] = \natural A$. Theorem 9 therefore yields as a corollary [3]'s main theorem (which is already a generalization of Brickell and Davenport's result [4]).

4 Characterizing Idealness

In the light of the previous section, IT-ideal (abbreviated *infra* as ideal) non-perfect schemes are naturally defined as those cases where formula (5) becomes an equality, that is :

Definition 11. A level-d non-perfect scheme $\mathcal{S}(P, \Sigma, S)$ is said to be *ideal* if

$$H(V_i) = \frac{\delta(i)}{d} H(S) \quad \text{for all } i \in P . \tag{6}$$

Note that several other (strictly equivalent) characterizations of ideal non-perfect SSS, such as

$$H(V_i) = \max_{X \subset P} I(V_i \,;\, S\,|V_X) \quad \text{or} \quad H(V_i) = \max_{X \subset P} I(V_i \,;\, SV_X) \qquad \forall i \in P$$

appear interesting but may require a more intricate theorization.

5 Ideal schemes and Matroids

Although theorem 9 states that K-ideal non-perfect schemes have a matroid structure, we will now show that, despite the light of the extended idealness concept given by definition 11, ideal non-perfect SSS do not always present this combinatorial morphology.

Let $\mathcal{S}(P, \Sigma, S)$ be a level-d ideal non-perfect SSS. Hence we have

$$H(V_i) = \frac{\delta(i)}{d} H(S) \quad \text{for all } i \in P$$

whereas theorem 7 yields,

$$\phi(A) = \frac{H(A)}{H(S)} \times d$$

as a rank function of a polymatroid on $Q = \{S_1, \cdots, S_d\} \cup P$. To upgrade $\mathcal{S}(P, \Sigma, S)$ as a matroid, conditions (2.3.4) and (2.3.5) of section 2.3 must also be fulfilled. ϕ's integer-valuation is guaranteed by the following lemma :

Lemma 12. *For all* $A \subset Q$

$$\phi(A) = \frac{H(A)}{H(S)} \times d \in \mathbb{N}.$$

Proof. Let $V_i = V_{i,1}||V_{i,2}||\cdots||V_{i,\delta(i)}$ with $H(V_{i,l}) = H(S)/d$ and define $\tilde{P} = \{(i,l) \mid i \in P, l = 1, \cdots, \delta(i)\}$. The function $\Delta : P \longmapsto 2^{\tilde{P}}$ defined by $\Delta(i) = \{(i,l)\}$ can be extended to 2^P as $\Delta(A) = \cup_{i \in A}\Delta(i)$. Using the access hierarchy $\{\Sigma_k\}_{k \leq d}$ on P, it is possible to construct a "shadow" access hierarchy $\{\tilde{\Sigma}_k\}_{k \leq d}$ on $\tilde{P}$ by $\tilde{\Sigma}_k^- = \Delta(\Sigma_k^-)$. $\mathcal{S}(\tilde{P}, \{\tilde{\Sigma}_k\}_{k \leq d}, S)$ is then a level-d non-perfect SSS such that $H(i,l) = H(S)/d$, which implies K-idealness. Consequently, for all $\tilde{A} \subset \tilde{Q} = \{S_1, \cdots, S_d\} \cup \tilde{P}$, $H(\tilde{A}) \times d/H(S) \in \mathbb{N}$ as a consequence of theorem 9 and in particular, for all $A \subset Q$, $\phi(A) = H(\Delta(A)) \times d/H(S)$ is integer-valued. $\square$
$\square$

Finally, condition (2.3.5) requires $\phi(i) \leq 1$ *i.e.* $\delta(i) \leq 1$ which means that

$$H(i) = \frac{H(S)}{d} \quad \text{for all } i \in P. \tag{7}$$

Naturally, we coincide *precisely* with Kurosawa *et al.*'s equation (3) on which was based K-idealness. In other words and against all odds, equation (7) only appears to be a *sufficient* matroidal condition and it is *a priori* unclear why (7) should be *necessary* ($\phi' \neq \phi$ may well define a matroid on Q for some ideal SSS that does not verify (7) at all).

Although a formal extension of the approach developped in section 3 does not seem trivial when $\delta(i) \neq 1$, negative existential evidence exists :

Let $p_1, \cdots, p_4$ be four large primes, set $p = \prod p_i$ and choose a secret $s \leq p$ such that $s \geq p/p_i$ for $i = 1, \cdots, 4$. Define on $P = \{1, 2, 3, 4\}$ the level-4 non-perfect scheme $\mathcal{S}(P, \Sigma, \{s\})$ by

$$
\begin{aligned}
v_1 &= s \bmod p_3 p_4 \\
v_2 &= s \bmod p_2 p_3 \\
v_3 &= s \bmod p_1 p_4 \\
v_4 &= s \bmod p_1 p_2
\end{aligned}
$$

The reconstruction process is simply a crt (as an exemple, if $A = \{1, 2\}$ then v_A is the only integer mod $p_2 p_3 p_4$ that fulfills :

$$
\begin{aligned}
v_A &= v_1 \bmod p_3 p_4 \\
v_A &= v_2 \bmod p_2 p_3
\end{aligned}
$$

and one can easily see that $v_A = s$ iff $A \in \Sigma_0$ where $\Sigma_0^- = \{\{1,4\}, \{2,3\}\}$). Despite (easy-to-prove) idealness, it appears that $B = 3$ and $A = \{1,4\}$ both belong to $G_{\mathcal{S}(P,\Sigma,\{s\})}$ but do not fulfill (2.3.8) which proves that $G_{\mathcal{S}(P,\Sigma,\{s\})}$ is not a matroid.

6 Open Questions

Although the results presented in this paper shed more light on non-perfect idealness, two challenging questions still persist :

- Characterize precisely the common morphology of ideal non-perfect schemes (which seems to lie somewhere between matroidal and polymatroidal structure).

- Characterize *ideal access hierarchies* for which there exist ideal non-perfect schemes having a matroidal morphology.

References

[1] J. Benaloh and J. Leichter, *Generalized secret sharing and monotone functions*, LNCS 403, Advances in Cryptology, Proceedings of Crypto'88, Springer Verlag, pp. 27–36, 1990.

[2] G. Blakley, *Safeguarding cryptographic keys*, Proceedings of AFIPS 1979 National Computer Conference, vol. 48, 1979, pp. 313–317.

[3] G. Blakley and G. Kabatianski, *On general perfect secret sharing schemes*, LNCS 963, Advances in Cryptology, Proceedings of Crypto'95, Springer Verlag, pp. 367–371, 1995

[4] E. Brickell and D. Davenport, *On the classification of ideal secret sharing schemes*, Journal of Cryptology, vol. 4, 1991, pp. 123–134.

[5] L. Csirmaz, *The size of a share must be large*, LNCS 950, Advances in Cryptology, Proceedings of Eurocrypt'94, Springer Verlag, pp. 13–22, 1994.

[6] Y. Desmedt, G. Di Crescenzo and M. Burmester, *Multiplicative non-abelian sharing schemes and their applications to threshold cryptography*, LNCS 917, Advances in Cryptology, Proceedings of Asiacrypt'94, Springer Verlag, pp. 21–32, 1994.

[7] S. Droste, *New results on visual cryptography*, LNCS 1109, Advances in Cryptology, Proceedings of Crypto'96, Springer Verlag, pp. 401–415,1996.

[8] W. Jackson, K. Martin and C. O'Keefe, *On sharing many secrets*, LNCS 917, Advances in Cryptology, Proceedings of Asiacrypt'94, Springer Verlag, pp. 42–54, 1994.

[9] M. Itoh, A. Saito and T. Nishizeki, *Secret sharing scheme realizing general access structure*, Proceedings of IEEE Globecom'87, Tokyo, pp. 99–102, 1987.

[10] E. Karnin, J. Green and M. Hellman, *On secret sharing systems*, IEEE Trans. Information Theory, vol. IT-29, no. 1, pp. 35–41, 1983.

[11] K. Kurosawa, K. Okada, K. Sakano, W. Ogata and S. Tsujii, *Nonperfect secret sharing schemes and matroids*, LNCS 765, Advances in Cryptology, Proceedings of Eurocrypt'93, Springer Verlag, p. 126–141.

[12] K. Okada and K. Kurosawa, *Lower bound on the size of shares of nonperfect secret sharing schemes*, LNCS 917, Advances in Cryptology, Proceedings of Asiacrypt'94, Springer-Verlag, pp. 33–41, Dec. 1994.

[13] A. Shamir, *How to share a secret*, Communications of the ACM, vol. 22, n.11, pp. 612–613, Nov. 1979.

[14] D. Stinson, *An explication of secret sharing schemes*, Designs, Codes and Cryptography, vol. 2, pp. 357–390, 1992.

Index

Lecture Notes in Computer Science

For information about Vols. 1–1280

please contact your bookseller or Springer-Verlag